CONTROLLED ENVIRONMENTS
for PLANT RESEARCH

CONTROLLED ENVIRONMENTS
for PLANT RESEARCH

ROBERT JACK DOWNS

COLUMBIA UNIVERSITY PRESS
NEW YORK AND LONDON 1975

Robert Jack Downs is director of the phytotron at the North Carolina Agricultural Experiment Station, North Carolina State University, in Raleigh.

Library of Congress Cataloging in Publication Data

Downs, Robert Jack, 1923–
    Controlled environments for plant research.

    Includes bibliographies and index.
    1. Growth cabinets and rooms.   2. Phytotron.
I. Title.
QK715.D68   1975   581'.07'24   74-20878
ISBN 0-231-03561-6

FOR ROSA JOY AND KATHLEEN

PREFACE

It would not be proper to enter into a discussion of controlled-environment facilities for plant growth without acknowledging at least some of the people who have contributed so heavily to our knowledge of the subject. F. W. Went is of course recognized as the originator of the phytotron. Less widely known are the efforts of P. J. Kramer and the contributions of the National Science Foundation, the Z. Smith Reynolds Foundation, and the tobacco companies toward continuation of controlled-environment research through the establishment of the Southeastern Plant Environment Laboratories at Duke and North Carolina State Universities.

Our ability to control the environment in plant structures has been enhanced by so many investigators in so many countries that they cannot be enumerated here. Major efforts have been made by L. G. Morris and his group in Silsoe, England, R. N. Morse and CSIRO engineering in Australia, T. Matsui and the Biotron Institute in Japan, as well as H. Miyayama, M. Konishi, and other members of the Committee for Environment Controlled Growth Rooms. Except for the now disbanded Phytoengineering Laboratory at Beltsville, headed by W. A. Bailey, no major research effort in this area has been made in the United States although there have been many individual contributions.

Proper use of controlled-environment facilities is possible largely because of the comprehensive research of the Physiology Laboratory at Beltsville. Although it too was disbanded by the USDA, this Beltsville group led by H. A. Borthwick, S. B. Hendricks, and M. W. Parker along with temporary members such as B. Cummins, L. T. Evans, H. Fredericq, H. Krug, H. Mohr, S. Nakayama, and Daphne Vince-Prue provided most of the information on phytochrome physiology that we use to control morphogenesis and reproduction.

Various instruments are shown in the figures, but no discrimination against other models and manufacturers is intended, and no guarantee of reliability is implied.

R. J. DOWNS

CONTENTS

Abbreviations   xi

Introduction   1

1. The Controlled-Environment Facility   5
2. Conditioning Systems for the Major
   Environmental Parameters   37
3. Environmental Measurements   87
4. Biological Aspects of Controlled-Environment
   Rooms   113
5. Specifications for the Plant-Growth
   Chamber   135
6. Testing and Maintenance of the Plant-Growth
   Chamber   153

Appendix 1. Conversion Factors   167
Appendix 2. Useful Constants for Water and Air   168
Appendix 3. Manufacturers of Growth Chambers
            in the United States   169
Appendix 4. Manufacturers of Growth Chambers
            outside the United States   170
Appendix 5. Types of Systems for Measuring
            Radiant Energy   171

Index   173

ABBREVIATIONS

alternating current	a-c
British Thermal Unit	Btu, BTU
calorie	cal
centimeter	cm
cubic feet per minute	cfm or ft^3/min or ft^3 min^{-1}
cubic foot	cu ft or ft^3
cubic meter	cu m or m^3
degree centigrade or Celsius	C
degree Fahrenheit	F
direct current	d-c
feet per minute	fpm or ft/min or ft min^{-1}
footcandle	ft-c
gallons per minute	gpm or gal/min or gal min^{-1}
gram	g
hectolux	hlx
high-energy reaction	HER
horsepower	hp
hour	hr
kilogram	kg
kilogram-calorie	kcal
kilometer	km
kilometer per hour	km per hr or km/hr or km hr^{-1}
liter	l.
lumen	lm
meter	m
microEinstein	μE
miles per hour	mph or mile/hr or miles hr^{-1}
nanometer	nm
photosynthetically active radiation	PAR
pound	lb
pounds per square inch	psi or lb/in^2 or lbs in^{-1}
square centimeter	sq cm or cm^2
square foot	sq ft or ft^2
square meter	sq m or m^2
watt	w
volts, alternating current	vac

INTRODUCTION

Environmental control for plant growth is often considered a modern development. Yet the environment was being controlled when man began to clear the land of its natural species and to plant crops in their place. The environment was controlled to a greater extent by eliminating competition from weeds, by applying manure to improve soil fertility, and by irrigating to provide unseasonal water.

Probably the first attempts at temperature control were made by the Romans, who used pits covered with mineral sheets (probably mica) and heated by fermenting manure for forcing vegetable and roses (Taft 1894). Some of the later Roman pits seem to have been heated with hot water circulated through bronze pipes. In England, grapes, oranges, and other fruits were adapted to the local climate by planting them against the south wall of masonry houses. The wall absorbed heat in the daytime and reradiated it at night to keep the plants warm enough to ripen fruit. Later the technique was extended by running the fireplace flues back and forth through the masonry wall to provide additional heat. Finally, of course, the south wall was made of glass and the plants moved indoors during the winter. The first such greenhouse in the United States was probably built in New York around 1764, and a few years later George Washington built one at Mount Vernon (Bailey et al. 1964).

The greenhouse enabled man to control the temperature so that plants could be grown throughout the winter. The possibility of profitable, year-round use of the greenhouse soon became apparent, and efforts were begun to ventilate and cool the structure as well as heat it.

It seems a small and logical step to go to completely controlled environmental conditions as typified by the plant-growth room. Early work with controlled conditions was limited chiefly to the control of root temperature (Jones, Johnson, and Dickson 1926). The limitations of the soil tanks illustrated the need to control simultaneously many factors of the environment. Although several earlier growth cabinets had been built, primarily for photoperiod control (Garner and Allard 1920, 1925), the first controlled-environment greenhouse and constant-condition room in the United States were probably those constructed at the Boyce Thompson Institute (Arthur 1928, Arthur, Guthrie, and Newell 1930). Many plant-growth chambers have been built since that time, each improving the ability to control one or more of the environmental factors. Usually these rooms have been designed and used by specialists investigating some aspect of environmental stress or environmentally induced plant response, such as photoperiodism.

Today the numbers of plant-growth chambers have increased to such an extent that one or more are present in nearly every laboratory engaged in plant research. Undoubtedly the future will see an even greater increase in the demand for controlled-environment facilities of all types (National Academy of Sciences 1966), and an increasing interest in temperature-controlled greenhouses is sure to be evident.

The research worker is asking questions of ever greater sophistication in his attempts to understand the systems that result in plant growth and development. Understanding of these physiological and biochemical systems provides the basic information for solving many applied agricultural problems. Unfortunately the attempts to analyze results of complex experiments often fail because the continuously fluctuating factors of the natural environment induce reactions that mask the responses to the experimental treatment. Thus the increased use of controlled-environment facilities is partly due to the realization that environmental control is a useful, and in fact increasingly necessary, tool for nearly all phases of basic and applied biological research.

A major stimulant to use of growth rooms is the availability of commercially designed, prefabricated models that can be installed with a minimum of effort and disturbance. Generally it is assumed that the commercial controlled-environment rooms avoids the disadvantages of built-in rooms and the uncertainty of owner-designed facilities. Unfortunately many investigators are dissatisfied with what appear to be mechanical instability and general unreliability of commercial chambers. Some scientists have even shut down their growth rooms and bitterly denounce the entire concept as useless until enough reliability can be engineered to enable completion of research projects without malfunctions. Frequently the researcher is also unhappy with the biological performance obtained with controlled-environment rooms. Granted that unbelievably bad engineering plays a large role in owner dissatisfaction, lack of information and experience on the part of the owner also contribute to unsatisfactory growth-room performance and to inadequate plant growth.

My purpose in this book is to describe the mechanical and biological systems encountered in modern controlled-environment facilities and to relate these systems to the problems of operation and plant growth. The discussion of the mechanical components of the growth chamber includes some of the frequent design problems; and in the sections on biological systems, the advantages of certain research techniques and cultural practices are pointed out. Introducing the biologist to the mechanical devices and showing the engineer some of the plant requirements will, I hope, help them to achieve better plant-growth-chamber design and operation.

References

Arthur, J. M. 1928. Artificial climate and plant growth. *Tech. Eng. News*, April 2–4.

Arthur, J. M., J. D. Guthrie, and J. M. Newell. 1930. Some effects of artificial climates on the growth and chemical composition of plants. *Amer. J. Bot.* 17:416–82.

Bailey, W. A., R. J. Downs, H. M. Cathey, and H. H. Klueter. 1964. Environment control in greenhouses. Paper no. 64-427, American Society of Agricultural Engineers meeting, St. Joseph, Mich.

Garner, W. W., and H. A. Allard. 1920. Flowering and fruiting of plants as controlled by the length of day. In *Yearbook of the U.S. Department of Agriculture*, pp. 377–400. Washington, D.C.: U.S. Department of Agriculture.

Garner, W. W., and H. A. Allard. 1925. Localization of the response in plants to relative length of day and night. *J. Agr. Res.* 31:555–66.

Jones, L. R., J. Johnson, and J. G. Dickson. 1926. Wisconsin studies upon the relation of soil temperature to plant disease. *Wisc. Agr. Exp. Sta. Res. Bull.* 71.

National Academy of Sciences. 1966. *The Plant Sciences Now and in the Coming Decade*. Washington, D.C.: National Academy of Science.

Taft, L. R. 1894. *Greenhouse Construction*. New York: Orange Judd Co.

FIG. 1.1. Interior of a 2.44 × 3.66-m plant-growth chamber with specular aluminum walls. Cool white fluorescent and incandescent lamps provide an illuminance of 460 hlx at pot level. Top-to-bottom air flow provides a temperature control of ± 0.25 C at the set point.

1 THE CONTROLLED-
ENVIRONMENT FACILITY

The term "plant-growth chamber" has been used to describe all types of controlled-environment facilities. Attempts to be more specific have resulted in a rough classification with descriptive names. *Plant-growth rooms* are artificially lighted rooms in which temperature, relative humidity, and sometimes other environmental factors are maintained at a fixed level or varied according to a predetermined program. The rooms are large enough to admit an operator and usually exceed 2.8 m² (fig. 1.1). Reach-in units of less than 2.8 m² are usually called *plant-growth cabinets* (fig. 1.2). Rooms or cabinets that provide a special environmental feature often have names that reflect it. *Dark rooms,* for example, may have all the environment-control capability of plant-growth rooms except light, which is deliberately excluded. *Cold rooms* may or may not be lighted and emphasize automatic defrost for low-temperature capability. *Germinators* are, of course, specially designed for seed germination studies and generally require precise temperature control and high relative humidity. *Dew chambers* are specifically designed to form and deposit dew on the plants.

Less elaborate plant growth chambers like *roomettes* are frequently quite useful. Roomettes are cabinets in which a ventilation system is used to maintain the same conditions as in the surrounding space. A roomette may contain an artificial light source or it may use an external source. It may also be equipped with humidity control and with heaters to raise the temperature, but the base-line conditions remain those of the area in which it is placed (fig. 1.3a). Roomettes function well as treatment chambers in studies of air pollution (fig. 1.3b) (Heck, Dunning, and Johnson 1968), phytochrome physiology, and CO_2 utilization. *Photoperiod rooms* are usually walk-in rooms that provide a low illuminance of about 1000 lux, often from a choice of several light sources. Temperature control may be obtained through use of unmodified ventilation air (fig. 1.4), or precise control over a wide range may be available by recirculating the air through a mechanical air-conditioning system.

FIG. 1.2. Reach-in 0.91 × 1.22-m controlled-environment cabinets with direct-expansion refrigeration.

Greenhouses should also be considered as plant-growth chambers. The degree of environmental control varies from simple heating to more elaborate year-round air conditioning and precise temperature control (fig. 1.5).

Combinations of environmental facilities are sometimes given special names. The best-known of these specialized facilities is the *phytotron:* a combination of temperature-controlled greenhouses and plant-growth chambers. Lang (1963) suggests that the first installation to deserve the name was a group of growth rooms built at the Kaiser-Wilhelm Institute in 1938. Unfortunately, they did not survive World War II, so their ultimate use and research capabilities cannot be evaluated. The original phytotron, the Earhart Laboratory at the California Institute of Technology (Went 1957), no longer exists, but others are in operation in Australia (Morse and Evans 1962), the Netherlands (Braak and Smeets 1956, Alberda 1958, Doorenbos 1964), Belgium (Bouillène and Bouillène-Walrand 1950), Sweden (Björkman et al. 1959, Wettstein 1967), Germany (Bretschneider-Herrmann 1962), Russia (Tumanov 1959a, 1959b), France (Chouard, Jacques, and Bildering 1972), Japan (Yamamoto 1972), and England (Hughes 1964). Additional phytotrons have

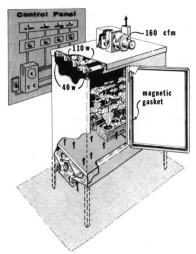

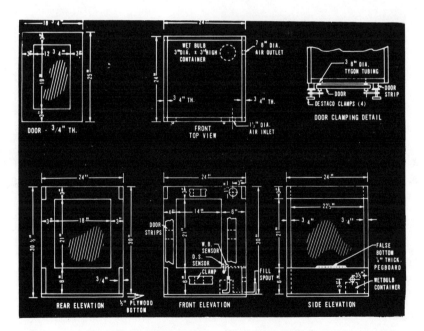

FIG. 1.3a.
Roomette designed for photoperiod and photomorphogenesis studies (Downs and Bailey 1967).

FIG. 1.3b. Roomette designed especially for air pollution research (Heck, Dunning, and Johnson 1968).

been reported in New Zealand (Climate Laboratory Operating Manual (1973) and Hungary (Rajki 1973), and a Biotron Institute has been developed in Japan (Matsui 1972). The Southeastern Plant Environment Laboratories in North Carolina consist of two phytotron units, one at Duke University in Durham and one at North Carolina State University in Raleigh (Kramer, Hellmers, and Downs 1970). An elaborate unit for both plants and animals—a biotron—is in operation in Wisconsin (Senn, Anderson, and Anderson 1965).

Phytotrons are usually laboratories expressly designed for studies of the response of plants to their environment, and they are so organized that many combinations of environmental factors can be studied simultaneously. Since all the controlled-environment systems are located in a building especially designed for that purpose, more attention can be given to maintenance than is usual with individual growth chambers. As a

FIG. 1.4. Photoperiod rooms used by H. A. Borthwick's Plant Physiology Laboratory (U.S. Department of Agriculture).

result, phytotron operation is more continuous, with fewer interruptions of the research programs than would be expected with individually owned units.

The plant-growth chamber is the basic unit or building block of phytotrons and biotrons. Therefore, a discussion of growth chambers singly or in small groups will be relevant to investigators planning more complex facilities as well as to the owner of a single chamber.

The plant-growth chamber is basically a simple, artificially lighted box in which temperature and other environmental factors are controlled. The apparent simplicity of the total system has led many designers into error. Standards used for "human comfort" air conditioning do not apply, and many of the techniques used in temperature control of buildings either cannot be used or must be considerably modified for use in the plant-growth chamber. Controlled-environment rooms (CERs) for

FIG. 1.5. Air-conditioned greenhouse constructed with small structural members and 1.22 × 2.44-m glass and using downward air flow (Southeastern Plant Environment Laboratories).

plants require precise control of temperature and relative humidity, and the established degree of control must be available over a wide range. The environmental conditions must be almost perfectly uniform through a plane of reference in which plants will be grown. Day and night conditions are usually different and often must go through a programmed rate of change. The light/dark cycles do not necessarily coincide with the natural sequence of day and night. The artificial light source produces the effect of a direct sun load throughout the room during the light period, and the load may abruptly change to nearly zero when the dark period begins. These problems are not insufferable barriers to designs that promote good growth, and many CERs work quite well when empty. When plants are introduced, however, they interfere with airflow and create unusual loads of latent heat. Latent heat is difficult to evaluate because it is not so easily detected as sensible heat. Sensible heat causes a change in the temperature of the surroundings and can be measured with a thermometer. Latent heat causes a change in the physical state of a substance; from a vapor to a liquid, for example. This change is accompanied by the production of heat, which must be removed by the temperature-controlling system. The factors that provide the water vapor, such as transpiration, evaporation from the substrate, frequency of watering and spillage, are difficult to translate into meaningful terms—if for no other reason than that plant species, pot size, and total leaf area are not constant. As a result, when plants are introduced the performance of a growth chamber almost always worsens.

Far too many expensive, poorly functioning growth chambers have been designed and built. The faulty engineering has resulted largely from a lack of appreciation by the designer of the complex combination of unusual, and often poorly defined, conditions that the equipment must provide. Generally engineers working through architectural firms have failed to provide satisfactory designs, and the more exalted their status the worse the result seems to be. They have no experience with such specialized jobs and try to fabricate a design based on off-the-shelf commercial components in the same manner as they design the heating and air conditioning of local schools. An equally large amount of poor design is the result of an amazing reluctance to benefit from the mistakes of others, combined with the interesting viewpoint that biologists must, by their very nature, be totally ignorant of engineering matters. It would therefore seem reasonable for anyone contemplating the design and construction of plant-growth facilities to obtain the services of a bioengineer or engineering biologist with experience in this specialized field.

Commercially designed and fabricated controlled-environment facilities should offer an excellent way to avoid many of the problems and uncertainties in the construction of plant-growth chambers. Unfortunately, comparison of performance characteristics from manufacturers' brochures about their various plant-growth chambers is virtually impossible. Attempts to evaluate chambers fail because there are no standards for measurement of environmental conditions and chamber performance. For example, only three of eight manufacturers report data on temperature uniformity, and only two report vertical temperature gradients. None describe the measuring system. Five of the eight companies do not include distance with their illuminance data, and none indicate the age of the lamps at the time the measurements were made. If the biologist is to obtain a unit that will serve his research requirements, he must write a set of comprehensive but reasonable specifications to ensure that the chamber will perform adequately. Asking the man who owns one to list his problems and the shortcomings of the design is an excellent first step.

The AIBS bioinstrumentation advisory council has prepared a set of guidelines for writing controlled-environment-room specifications (Busser 1971). Unfortunately, they do not explain the choices involved or supply background material on which to base the specifications. One may safely assume that no improvements will be made unless the prospective owner demands them; and improvements are not always costly although, prior to bidding, many manufacturers will claim great cost increases. The prospective owner will receive exactly and only what is in the specifications. Exactly what they literally say, and even that may be altered by interpretations if the statements are not clear and concise. A knowledge of the important component parts of the plant-growth chamber and their function and interaction is definitely required for the successful design. It is equally important for the purchase and use of such equipment.

Basically the plant growth chamber consists of (1) a refrigeration system which controls temperature and relative humidity in (2) a structure that has (3) a natural or artificial light source.

Refrigeration System

Prefabricated plant-growth chambers generally use a direct-expansion, mechanical refrigeration system. Large walk-in rooms and complexes of rooms sometimes use a secondary-coolant system in which the basic, direct-expansion unit is used to cool a liquid such as aqueous ethylene glycol. The secondary coolant is then used to remove heat from the growth chamber without a change of state.

A mechanical refrigeration system depends on recirculation of a primary refrigerant that changes from gaseous to liquid to gaseous phases. Such systems are composed of four main parts: expansion valve or other pressure-reducing device, evaporator, compressor, and condensor (fig. 1.6). Various auxiliary devices such as receivers, surge tanks, driers, valves, and controls make up the final design (*ASHRAE Guide and Data Book: Fundamentals and Equipment* 1965; *ASHRAE Handbook: 1973 Systems* 1973). When all the components are assembled, they might be illustrated by the schematic drawing in figure 1.7.

Pressure Reducer

Some kind of device must be incorporated into the system to reduce the pressure of the liquid coming from the condenser and to discharge a mixture of low-temperature, low-pressure liquid into the evaporator. Both expansion valves and capillaries are used on plant-growth-chamber refrigeration systems, with thermostatic expansion valves being the most common (fig. 1.8). The thermal expansion valves automatically adjust the amount of liquid admitted to the coil as a function of the load and permit several evaporators to be operated in parallel to a single compressor.

Evaporator

The evaporator or cooling coil is a heat exchanger in which the refrigerant absorbs heat from the conditioned space (fig. 1.9). The refrigerant is discharged from the cooling coil as superheated vapor (temperature higher than the saturation temperature corresponding to the pressure) at moderately high temperature and low pressure.

Sizing the evaporator to meet the requirements of the system is one of the perplexing problems of growth-chamber design, because a

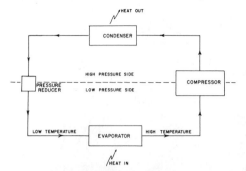

FIG. 1.6. Simplified block diagram of the main components of a refrigeration system.

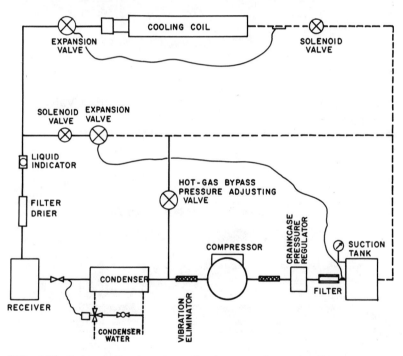

FIG. 1.7. Schematic drawing of a hot-gas bypass, direct-expansion system similar to that used on many plant-growth chambers.

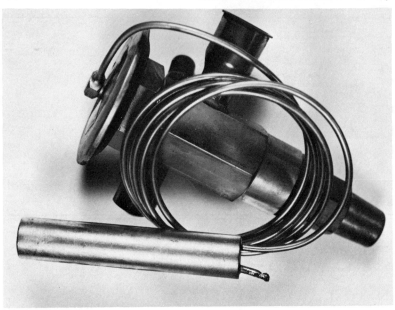

FIG. 1.8. A thermostatic expansion valve (Sporlan Valve Co., St. Louis, Mo.).

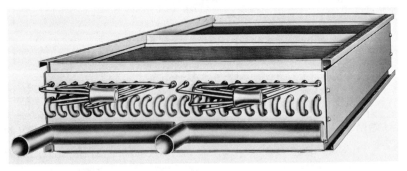

FIG. 1.9. Typical, multicircuited evaporator coil which may have from 2 to 22 parallel circuits fed from one expansion valve (*Trane Air Conditioning Manual* 1968).

cooling coil is usually selected for a critical design condition. Each change in the set point for temperature or relative humidity therefore establishes a different set of conditions between coil and compressor performance. For example, a coil selected for 25 C and 85% RH might be rated at over 2 tons/ft² surface, but when it was operated with inlet air of 20 C and 85% RH, the rating would be about half that amount. Generally the farther the set point deviates from the design point, the poorer the coil performance. However, by careful selection of cooling coils and a good deal of experience, the growth-chamber manufacturer and the experienced designer are able to provide satisfactory temperature control over a range of about 28 C. Since a definite range limit exists, the range should be shifted rather than extended when very high or very low temperatures are required.

Compressor

The compressor (fig. 1.10) is a device for compressing moderately high-temperature, low-pressure vapor from the evaporator into a high-pressure, high-temperature vapor. Reciprocating compressors are most commonly used with plant-growth chambers. Usually these are her-

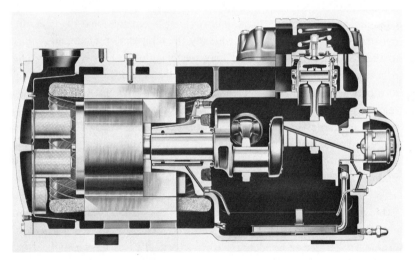

FIG. 1.10. Cross-sectional view of a hermetic compressor (*Trane Air Conditioning Manual* 1968).

metic units with the motor and compressor sealed together in a container so that the motor operates in a refrigerant atmosphere. Hermetic compressors usually make less noise than open ones.

Condenser

The function of the condenser is to remove the heat absorbed by the expanding gas in the evaporator and the heat gained in the compressor. Upon removal of heat, the vapor is converted to the liquid phase and is available for reexpansion. Condensers may be water-cooled or air-cooled, and commercial chambers are available with either type. Since operational problems can frequently be traced to the improper use of a particular type of condenser, it seems appropriate to discuss the various types and their limitations.

Air-cooled condensers. Air-cooled condensers can be broadly classified into chassis-mounted and remote types. The chassis-mounted condenser is usually installed on a common base plate with the compressor. Since the condenser is a sink for heat absorbed in the evaporator, the warm air discharge will obviously heat up the air surrounding the growth room. Unfortunately, growth chambers are often located in poorly ventilated areas such as storerooms, hallways, or basements, and the heat output may be considerable. Consequently the temperature of the air supply to the condenser increases and may reach undesirable levels. The total heat rejection of a condenser is proportional to the difference in temperature between entering air and the compressed refrigerant vapor. As heat rejection falls below the design level the heat extracting capability of the condenser (the effective condenser size) decreases, with a corresponding decrease in the net refrigeration effect. Therefore an otherwise adequate refrigeration capacity could become inadequate, and the biologist would find that growth-room temperatures could no longer be maintained. I have seen installations where the discharge air from the condenser of one room is blown directly into the suction of the condenser fan of a second room, and that condenser is discharged into the suction of the fan on a third room. The effective condenser size is markedly less as we move from room one to room two to room three.

Plant-growth rooms are sometimes installed in unheated hallways or buildings, or a remote-type condenser may be mounted outdoors to avoid raising indoor temperatures. In winter the effective condenser capacity increases owing to the low temperature of the entering air. When the ambient temperature falls below about 5 C, however, the head pressure (the operating pressure measured in the discharge line of the compressor outlet) decreases to an abnormally low level. The result is an insufficient pressure across the expansion valve and an inadequate amount of refrigerant delivered to the evaporator. When a low-pressure switch controls compressor operation, the switch may not be activated because the pressure corresponding to the air temperature at the condenser is below the pressure setting of the switch, and the compressor will not start. During off-cycles a low outdoor ambient temperature may cause the refrigerant to migrate from the evaporator and receiver and condense in the cold condenser. The system pressure then drops and, upon starting, an inadequate amount of liquid refrigerant going to the evaporator results in compressor cycling until the suction pressure (operating pressure at the compressor inlet) reaches an operating level above the setting of the low-pressure switch. If the outdoor temperature should drop still further, the system pressure may be reduced so far below the operating point of the low-pressure switch that the compressor will not run. On-off cycling of the condenser fan for low-temperature operation has not proven successful because it creates large fluctuations in head and evaporator pressures. Multispeed fans or dampers to divert the air stream seem to work fairly well, as does bypassing part of the refrigerant vapor from the compressor directly to the liquid receiver.

Air-cooled condensers operate at a higher head pressure than water-cooled ones, reducing the compressor capacity and increasing the power requirement. Thus a 3-hp air-cooled unit would be required to do the same job as a 2-hp water-cooled one.

Air-cooled condensers require at least 9.3 m² of finned coil surface per compressor horsepower, and the fan must deliver at least 14 m³/min per compressor horsepower (King 1971).

Water-cooled condensers. Water-cooled condensers (fig. 1.11) require water at the rate of approximately $G \Delta t = 63$; where G is the flow in liters per minute per ton and Δt is the rise in temperature of the water as it passes through the condenser. Water-cooled condensers may use city water or evaporatively cooled water from spray ponds or water towers.

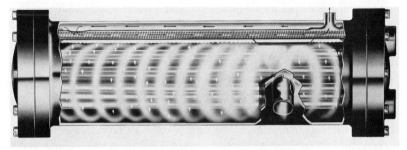

FIG. 1.11. Cross-sectional view of a shell and tube condenser (*Trane Air Conditioning Manual* 1968).

With city water a temperature rise of 11–14 C seems to be commonly used for design, so at least 4.0 to 5.7 l. min⁻¹ ton⁻¹ of water would be required. We are referring here of course to a ton of refrigeration as used in American commerce. It is equal to 12,000 Btu/hr or 3,024 kcal/hr. Since use of city water usually means a one-pass system, these condensers waste water in large amounts and cannot be recommended even when state and municipal codes are not prohibitive.

With a water-tower design, conditions may call for a 2–3.5 C temperature rise. The smaller temperature rise provides a somewhat better head pressure and better efficiency. Required circulation rates would be 18–30 l. min⁻¹ ton⁻¹. Since the latent heat of water is about 555 cal/g, evaporating 1 kg of water would be expected to cool nearly 100 kg of water about 5.6 C. In this case only about 1% of the water would be lost by evaporation while it provided 5.6 C of cooling. Clearly, evaporative cooling conserves water.

The evaporative cooling system is customarily located outdoors and is subject to the same cold-weather problems as the remote air-cooled condenser, plus the fact that the water may freeze. Condensing-water regulating valves are usually installed on larger systems and sometimes on small ones. The purpose of these valves is to keep the condensing pressure at a point that will load the compressor but not overload it. Two-way valves are not recommended because in cold weather they reduce the system's total flow. With three-way regulators, total flows are maintained while the supply to each condenser is modulated.

When cooling towers are to be operated in freezing weather, a further control is needed in addition to the condenser-water regulators. Water-tower fans can be modulated with some success, but the best method is to use a thermostatically controlled, three-way proportioning valve to recycle some of the warm return water to mix with the colder water supplied by the tower. With light loads and cold weather the tower can be entirely bypassed, so an indoor sump may be required.

To summarize the refrigeration cycle: Liquid refrigerant slightly above room temperature travels from a storage area or receiver to a metering device such as an expansion valve. The expansion valve reduces the pressure of the refrigerant as it enters the evaporator coil. As a result of pressure reductions the refrigerant is also at a lower temperature. As heat from the area is absorbed by the evaporator and consequently by the refrigerant, more and more of the liquid boils off and changes to a gas. By the time the refrigerant reaches the outlet of the evaporator it has all been vaporized and is in a gaseous state. The refrigerant vapor is drawn into the compressor by the action of the much lower pressure created in the "low" side of the unit, the suction side. The low-temperature, low-pressure refrigerant is then compressed into a high-pressure, high-temperature vapor. Passing into the condenser, the vapor is cooled by air or water and condenses or liquifies. The liquid refrigerant flows to the receiver, and the cycle is ready to begin again.

The receiver does not contribute to the refrigeration cycle but is generally a necessary part of the total system. A strainer is usually placed in the liquid line before the expansion valve to keep dirt or scale from clogging it. Sometimes one also finds a strainer on the suction line to protect the compressor from dirt and scale. A dehydrator or drier containing a chemical absorbant or desiccator should definitely be used, and a sight glass to indicate refrigerant levels is most helpful (fig. 1.12). A complete direct-expansion system similar to those used on many growth chambers is shown in figure 1.7.

Sizing the Refrigeration System

Exact calculation of the refrigeration requirements of a controlled-environment room for plants is complicated by a number of unknowns, some of which have already been mentioned. A reasonable approximation, however, is relatively easy to calculate and is probably adequate. For an example of cooling load calculations let us assume that a chamber measuring 8×12 ft (2.44×3.66 m) with the usual 7-ft (2.13-m) inside height is to be placed in a building that is air conditioned to the usual 75 F (24 C) and 55% relative humidity level. We will also assume that the lowest chamber temperature will be 45 F (7 C). (Because the air-conditioning engineers in this country still use the foot-pound-second system in their calculations, and probably will continue to do so for some time, we have used it here, followed by the more desirable metric units in parentheses. We have also used abbreviations such as cfm for cubic feet per minute and fpm for feet per minute because they are an integral part of American engineering parlance.)

Heat transmission through the walls, Q_w, is a function of the total area, A, the difference between temperatures outside, t_o, and inside, t_i, and the coefficient of heat transfer, U. The value of U will depend on the wall material, wind velocity, and radiant energy. If the walls are aluminum-styrofoam sandwich panels, the manufacturer would give a U-

FIG. 1.12. Refrigerant driers (Sporlan Valve Co., St. Louis, Mo., and Henry Valve Co., Melrose Park, Ill.) and typical sight glass (Henry Valve Co.).

valve of 0.12 Btu hr^{-1} ft^{-2} °F^{-1} (0.58 kcal hr^{-1} m^{-2} °C^{-1}) to a 2-in. (5-cm) thickness and a value of 0.08 (0.39) to a 3-in. (7.6-cm) thickness. Since our room has a total wall area of $56 + 56 + 84 + 84 = 280$ ft^2 (26 m^2) and $\Delta t = 75 - 45 = 30$ F (17 C), losses from the 2-in. (5-cm) thick walls would total 1008 Btu hr^{-1} (254 kcal hr^{-1}). Floor losses depend on additional factors but would be at least 346 Btu hr^{-1}.

Actual measurements made with an inside air temperature of 10 C during the dark period show that the inside wall temperature of a 5-cm panel is 10.5 C and the outside wall is 20 C when the outside air temperature is 24 C. The outer wall temperature will continue to decrease as the interior temperature is reduced. If the lights are on, the inner wall will warm as much as 8 C and the outer wall will approach the temperature of the outside air and become about 23 C.

The cooling load due to radiant energy is a function of lamp type, total installed watts and the transmission characteristics of the barrier. The lighting system of the 2.44×3.66 m room may consist of 84 fluorescent lamps of 215 w each, and 4,800 w of incandescent lamps. The infrared output of the two lamp types is 36 and 72%, respectively, and the visible output is 22 and 10% of the initial watts (fig. 2.19).

Methyl methacrylate plastic apparently transmits about 90% of the infrared out to 1500 nanometers (nm) and about 92% of the visible light. Thus the lighting load would consist of $(18,060 \text{ w} \times 0.36) + (4,800 \text{ w} \times 0.72) \times 0.90 = 8,961$ w of infrared and $(18,060 \text{ w} \times 0.22) + (4,800 \times 0.10) \times 0.92 = 4,097$ w of visible energy; a total of 13,058 w or 44,528 Btu hr^{-1} 11,221 kcal hr^{-1}).

Conduction through the 96 ft^2 of barrier, $Q_b = AUt$, to a room operating at 75 F (24 C) might be based on a lamp-loft temperature of 105 F (40 C). Given manufacturer's data for U of 1.09 (5.3), conduction would increase the load by 3,139 Btu hr^{-1} (791 kcal hr^{-1}). Since there is a substantial air flow on both sides of the barrier, this U-factor is more probably of the order of 1.40–1.50 (6.8–9.9 kcal hr^{-1} m^{-2} °C^{-1}), a situation that would increase the conduction load to 4,320 Btu hr^{-1} (1,088 kcal hr^{-1}). Total sensible heat load would thus be $1,008 + 346 + 44,528 + 4,320 = 50,202$ Btu hr^{-1} (12,650 kcal hr^{-1}).

Latent heat would depend on the amount of water to be removed from the room. Evapotranspiration and spillage make the most significant contributions to this water. If we assume that 3 lb (1.4 kg) of water must be removed per hour from the growing space, then the latent heat load would be 3×1076 or 3,228 Btu hr^{-1} (813 kcal hr^{-1}).

Many growth chambers are equipped with a fan to move outside air into the conditioned space. This so-called makeup air could create a significant load increase, especially if applied at the recommended rate of 5 cfm/ft^2 (1.5 m^3 min^{-1} m^{-2}) (Morse 1963). For our example room, the makeup air would be 30 F (17 C) higher than the lowest room temperature; at 5 cfm/ft^2 (1.5 m^3 min^{-1} m^{-2}), $Q_a = 30 \times 1.08 \times (96 \times 5)$ or 15,552 Btu hr^{-1} (3,919 kcal hr^{-1}).

The latent load of the makeup air would be calculated as cfm $\times 60 \times 0.075 \times 1076 \times (w_o - w_i)$ and would amount to nearly 1,557 Btu hr^{-1} (3,924 kcal hr^{-1}); where 60 is a conversion from minutes to hours, 0.075 is standard air density in lb ft^{-3}, 1076 is the Btu released in condensing 1 lb of water vapor from the air, and w is the humidity ratio in pounds of moisture per pound of dry air. The w_o for outside air at 75 F and 55% R.H. would be 0.0120, and w_i for inside air at 45 F and 55 R.H. would be 0.00350.

The total load without makeup air would therefore be about 50,000 Btu hr^{-1} (12,600 kcal hr^{-1}) or 4.2 tons. Including the makeup-air load of 31,123 Btu hr^{-1} (7,842 kcal hr^{-1}), the total load would become 81,123 Btu hr^{-1} (20,443 kcal hr^{-1}) or 6.8 tons.

Pull-down tonnage, the refrigeration capacity needed to decrease the temperature by a certain number of degrees in a specified time, is not included. To calculate pull-down requirements, the total weights of chamber components plus pot and substrate weight are multiplied by the various specific heats times the temperature change desired in 1 hour. Water in the substrate would have to be included, of course. A 10-F (5.6-C) pull-down in 20–30 min could easily result in a 2–2.5-ton increase in the required refrigeration capacity.

Specialized Controlled-Environment Facilities

The problems of design and maintenance of specialized facilities are unusual enough to warrant discussion of a few types, such as the seed germinator and the inoculation chamber.

Seed Germinators

The design requirements of seed germinators include precise temperature control over a wide range and a saturated humidity level at all temperatures. Sometimes light is also a requirement, and while the intensities need not be high, the distribution must be uniform over each tray or shelf. Commercial incubators almost always fail to meet the design requirements, and as Issacs, Shenberger, and Carter (1952) reported, a number of commercial seed germinators also fail to maintain sufficiently high humidities to prevent premature drying of the substrate.

The simplest form of seed germinator heats and cools by conduction rather than convection. A recirculated cold-water supply is passed through tubing mounted on the walls. The rate of flow, which determines the temperature, is controlled by manual adjustment of a valve. A large evaporative surface keeps the relative humidity high. The humidifier may be a thermostatically controlled heater in a pan of water that covers the entire bottom of the chamber. When coil and water temperatures are matched, such germinators provide good temperature control and high relative humidity with a very low maintenance factor.

More elaborate germinators may utilize a spray of water directed against one or more of the walls in addition to the water coils (Verhey 1955, Nelson, Wolf, and Stone 1963). Instead of using a single thermostat in the air, germinators like those designed by Nelson, Wolf, and Stone (1963) may use a coiled-capillary hydraulic thermostat in the chamber water reservoir with an air temperature override to allow the heater or cooling water to operate until the setpoint is reached.

Cycling or alternating temperatures are often required in seed germinators because many kinds of seeds germinate poorly in a constant-temperature condition. A rapid temperature change is usually considered desirable, with a 10-C alternation between 25 and 15 C or 30 and 20 C the most common. The controls of the germinator designed by Nelson, Wolf, and Stone (1963) are particularly suitable for alternating temperatures. Commercial two-temperature germinators of the conduction type have larger heaters and two sets of coils and thermostats, one set for each temperature.

Commercially made seed germinators work very well. Only two

major problems occur. These germinators have a tendency to leak water, which is not only messy but can provoke a very short heater life. As in so many controlled-environment facilities, the controls of many commercial germinators need to be replaced with better, more reliable types.

In convection germinators, where heating and cooling require the use of fans, other types of humidification systems such as spray chambers could be used. Hinkle, Spillman, and Shenberger (1968) describe an ejector–venturi scrubber saturation system that overcomes some of the problems of maintaining a saturated atmosphere in an alternating-temperature germinator. This design has no moving parts—a marked advantage in high humidity atmospheres. Most moving parts in the convection germinators, such as fan motors and bearings, should be mounted outside to reduce maintenance problems.

Lighting in germinators is usually bad. Uniform lighting can be attained but only through the sacrifice of shelf space. A 12-tray dark germinator thus would become a 3-tray one if lighted properly. The high humidity can create problems in the lighting system through corroded lamp holders and other parts, and the temperature differences between germinators can cause large variations in light output if fluorescent lamps are used.

The cooling system for seed germinators is usually a secondary-coolant system in which the direct-expansion compressor chills water in an evaporator or sump. Except for the precaution of filtering the water of debris picked up in open-pan germinators, one need expect no special design or maintenance problems beyond those normally associated with chilled water systems.

Inoculation Chambers

Studies involving plant pathogens often require some method of keeping moisture on the plant surfaces for periods of 16 hr or more. The commercial dew chamber performs this task quite well, but the small volume of the interior space limits the number and size of the plants to be inoculated. In the dew chamber a hot water evaporator with a large surface is used to raise the relative humidity. The walls and ceiling of the chamber are chilled and become a heat sink to which the plants radiate.

Ordinary plant-growth rooms have been adapted to serve as inoculation rooms by installation of horizontal-fan nozzles that spray de-ionized water into the atmosphere. Very fine droplets are obtained by driving the water through the nozzles with compressed air at 3.5–5 kg/cm². The on-off schedule depends upon plant type and air temperature as well as water temperature. In our laboratories, once the room was initially saturated at 25 C, a schedule of 5 min on in every 20 min kept corn leaves wet without appreciable runoff.

Commercial mist chambers as large as 1.2 × 2.4 m have been made on the same principle using mist nozzles driven by compressed air. These units have temperature control and lighting systems.

The Greenhouse

Any discussion of greenhouse environment control can profitably begin with a quote from Lawrence (1957):

The glasshouse seems a simple structure; achieving some control of air temperature by heating or ventilating seems a simple matter and meeting the requirements of good experimentation seems a simple procedure. All this is in fact true when one knows what to do, but it is also a fact that in many research establishments glasshouses have been erected with little understanding of the factors involved and with a simplicity of outlook which can only be described as naive. I mention this because we cannot profitably discuss control of glasshouse environment unless we first appreciate the widespread lack of it.

A visit to a few greenhouses around the country makes me wonder if Lawrence has not understated the problem. I have seen greenhouses with pipes and conduits running back and forth overhead, artificial light systems with huge solid reflectors, or fluorescent tubes so close together that a slat-house effect results. Heating and ventilating controls often border on the ridiculous; sometimes greenhouse wet-pad systems are even laid out in series so that the discharge air of one house becomes the inlet supply to the next.

Clumsily built greenhouses may be adequate if the aim is only to ensure survival of cold-sensitive plants during winter, but for commercial production or research purposes the greenhouse should be designed by experts. From what I have seen, whoever is doing the engineering of many of these houses does not qualify.

Structure

Greenhouse structural design is rather standardized, and to attempt changes can prove difficult and costly. In most areas of the country, greenhouse design conforms to general building codes. While these codes may have prevented extensive structural failures, they have also resulted in considerably overdesigned structures. The greenhouse frame must of course resist the various loads imposed upon it while simultaneously intercepting the minimum amount of light. In order of decreasing magnitude, the greenhouse loads are wind, snow, any crop load suspended from the structure, and the dead weight of glazing and structural members. According to Morris and Johnson (1965), 15 kg for each square meter of floor area should be allowed for crop loads suspended from the structure. In research houses, where additional, originally unintended equipment is invariably added, the structural dead load should probably be doubled. For example, the addition of one high-intensity-discharge lamp fixture can place a 32-kg load on a single point of the structure.

Structural members cast dense shadows that remain for an appreciable period of time. Even when very large glass panels are used with the smallest possible structural members, shadowing is severe (fig. 1.13). Considerable research still is needed to determine the influence of glass and structural member size, number, and spacing on the amount of light reaching the plant material.

Glazing

The relative merits of glass and plastic coverings seem to depend on the function of the structure. With both types of glazing, transparent and diffusing materials are available. According to Seeman (1952), the diffusing surface reduces, in fact virtually eliminates, the large variations of light intensity caused by the opaque parts of the structure. Most of the manufacturers claim (and Seeman agrees) that there is little reduction in the total intensity. Since it is difficult to reconcile such statements with what the eye appears to detect, we took some measurements to compare lightly diffusing with clear plastic. When the plastic was almost directly on top the photocell, there was 95% transmission through the clear plastic, versus 85% through the translucent. However, if the material was

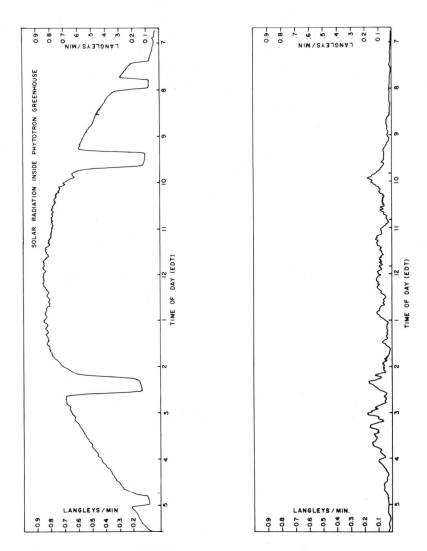

FIG. 1.13. Daily light intensity in a greenhouse at Southeastern Plant Environment Laboratories. Note the severe shading effects caused by the small structural members. Upper curve: clear day, full sun; lower curve; complete overcast.

located 10 cm above the photocell (with no light leakage from the sides), the diffuse plastic transmitted only 79% whereas no change occurred in the transmittance of the clear material. In a spectrophotometer the same diffusing plastic had 25–30% transmission when the clear was adjusted to 100%. The only safe conclusion seems to be that the amount of light lost by use of diffuse glazing depends on the way the measurements are made. Whether the plants finally intercept the light that apparently escapes the measuring device is currently debatable. In the plant-growth chamber, use of diffusing plastics for barriers may give different results in both measurement and plant growth because of the integrating effect of the highly reflective walls. Plastic and fiberglass glazing all suffer to a greater or lesser degree from weathering, which certainly reduces the transmittance and can alter the spectral distribution. The initial cost of plastic glazing is usually lower, however, and some types have greater hail resistance than glass.

Resistance to change on the part of many greenhouse manufacturers usually forces the use of standard greenhouses with small glass panes. Even our attempts to obtain greenhouses with large Dutch glass (71×142 cm), commonly used in Europe, met with resistance in the form of considerably higher cost.

Orientation

As one approaches the equator, greenhouse orientation becomes less important because orientation effects are lower at solar angles of 0–40° than at angles of more than 40°. East-west orientation at latitudes above 35° generally allows more light into the greenhouse than does a north-south alignment (Lawrence 1957, Whittle and Lawrence 1959). This advantage exists between October and March, when the sun angle is low. Between March and October, however, a greater light intensity is obtained with greenhouses oriented north-south (Nisen 1965). Moreover, Stocker (1949) claims the north-south alignment is more natural as far as daily CO_2 assimilation is concerned.

Heating

The approximate heat loss from a greenhouse can be calculated from $H = AU \ (t_i - t_o)$, where A is the total glass area and U is the coefficient of transmission of glass. The maximum inside temperature desired is t_i, and t_o is the outside design condition. Additional losses occur as a result of infiltration and wind velocity. The amount of heat needed to warm outside air entering by infiltration would be $H = 0.240V\rho(t_i - t_o)$, where 0.240 is the specific heat of air in Btu/lb, V is the volume of infiltration air, and ρ is the air density at t_o. For example, if the ventilators of a 100×28-ft greenhouse fail to seal tightly, a crack 0.016 in. $\times 400$ ft may result. In a 15-mph (24-km/hr) wind the loss would be 75 ft^3 per foot of crack (7 m^3/m); if $t_i - t_o = 65$ F, then

$$0.016 \times 75 \times 400 \times (t_i - t_o) = 35,100 \text{ Btu/hr (8,845 kcal/hr)}$$

or about the same as exchanging one air volume per hour to the outside:

$$0.240(28,000 \text{ ft}^3/\text{hr}) \times 0.08(t_i - t_o) = 34,944 \text{ Btu/hr}$$
$$(8,806 \text{ kcal/hr})$$

The effect of wind on U values would be different for roof and side glass, as would the effect of wind on infiltration. The wind factor can, however, be approximated as 1.10 for a 15-mph wind, with about a 4% increase for each additional mile per hour. Tables have been worked out from which a rapid approximation of greenhouse heat losses can be obtained (How to calculate greenhouse heat loss 1962).

Greenhouse heating systems have been installed in many different ways with varying degrees of success. Generally, however, finned pipe containing hot water or steam should be placed along the outer walls. In addition some means of overhead heating should be provided, preferably a system like the fan and tube. With this method, air from a heat exchanger is blown through a thin, perforated plastic tube which is kept inflated by air pressure.

Ventilation and Cooling

The methods now prevalent for greenhouse cooling and ventilating use some form of evaporation. The most popular method, the wet pad and fan system, uses fans on one side of the greenhouse to pull air in through wed pads on the other side of the house. Water trickles down through the pads and is collected at the bottom and returned to a sump for redistribution (fig. 1.14). Standards for this type of system have been established by the National Greenhouse Manufacturers Association (Standards for ventilating and cooling greenhouses 1962).

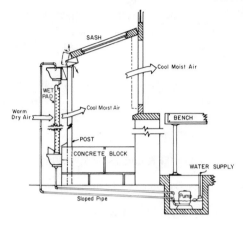

FIG. 1.14. Schematic drawing of a wet-pad evaporative cooling system (Bailey 1965).

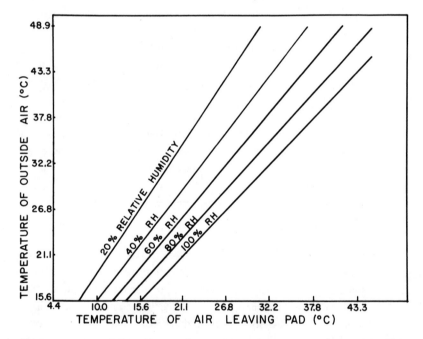

FIG. 1.15. Effects of outdoor relative humidity and air temperature on the processed air leaving the wet pad (Walker and Abernathie 1964).

Bailey (1965) shows that the maximum cooling with recycled water is obtained with a velocity of 150 fpm (46 m/min) at the pads. Fans should of course be numerous enough to provide uniform air flow across the greenhouse, and controls should be selected and installed carefully. The sensing elements should be placed in shielded, aspirated housings located near the midpoint of the greenhouse at plant level.

The efficiency of this type of cooling system is measured as a percentage of dry-bulb temperature change (the temperature indicated by a thermometer after correction for radiation). The potential cooling is shown in figure 1.15. For any outdoor temperature, the lower the relative humidity the greater the cooling obtained. Data taken at Beltsville, Maryland (Bailey 1965) show that maximum cooling occurred at the time each day when relative humidity was lowest (fig. 1.16).

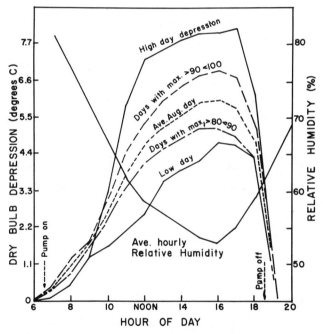

FIG. 1.16. Dry-bulb depression of air passing through a wet pad installed on a greenhouse at Beltsville, Maryland (Bailey 1965).

Shading. The oldest method of temperature control for the green-house is reduction of the insolation by shading with whitewash or other white paint-like material. The disadvantage of this approach is that the material is rather permanent, remaining in place throughout the summer. Wooden lath or plastic screen blinds can be unrolled over the greenhouse on each sunny day and left rolled up on cloudy days.

A film of water running over the greenhouse roof is much too thin to absorb any appreciable amount of solar radiant energy, although it will conduct away the heat absorbed by the glass and structural members. If the opacity of the water is increased, however, by adding a dye especially developed to absorb solar radiation, this should prove an effective shading method. Canham (1965) reports that Solivap Green is just such a dye and that it effectively removes 54% of the solar radiant heat.

Covers for the greenhouse to be used at night during cold months could reduce heat requirements by 25%. Double glass with a 12-mm air space would reduce the U value, and presumably the heating requirements, by 50%.

Refrigerated air conditioning. Really uniform temperatures and close control about the set point can only be obtained with refrigerated air conditioning. Air flows are uniform, and full sunlight can be utilized since shading is unnecessary.

The air-conditioned greenhouse should be designed in much the same way, based on the same principles and reasoning, as the controlled-environment room. For example, if 30-m/min air flows are considered optimum in the plant-growth room, why should 7 m/min be acceptable for a greenhouse? The greater the volume and the larger the temperature difference between inlet and discharge air, the greater the temperature variation in the space. For a given cooling load, the air exchange rate and the $t_o - t_i$ values are related as

$$\text{cfm} = \frac{Q_s}{1.08(t_o - t_i)}$$

where Q_s is the sensible heat of the space. Therefore a 22×24-ft lean-to greenhouse with a calculated sensible heat load of 140,000 Btu/hr would require

$$\frac{140,000}{1.08 \times 10} = 12{,}962 \text{ cfm } (366 \text{ m}^3/\text{min})$$

which is barely the minimum for good temperature control. A much more satisfactory Δt of 4 F (2.2 C) would require 32,400 cfm (916 m³/min), or about five air changes per minute, and 61-fpm (19-m/min) velocity across the plant leaves. A 100-fpm (30-m/min) velocity would require 52,800 cfm (1,494 m³/min), or nine air changes per minute, and provide a Δt of 2.5 F. These last rates are out of the realm of practicality; the five or six changes per minute provide excellent temperature control and uniformity, and the air flow seems adequate for good gas exchange between leaf and atmosphere.

Sizing the refrigeration system. The design outdoor conditions of the area, the minimum greenhouse temperatures desired, and the maximum solar radiation are the major factors in determining the sensible heat load of a greenhouse. For example, the top of a south-facing lean-to greenhouse might be calculated to have a solar load of 284 Btu/ft², whereas the south wall might have about 100 Btu/ft². Heat conducted through the glass would be added, of course, as $H = AU(t_i - t_o)$.

In a greenhouse full of plants the latent heat load can be appreciable and must not be left out of the calculations. Some engineers have assumed that moisture evaporation requires as much as half of the sensible load. The total refrigeration requirements for greenhouses in the United States have been calculated as high as 0.2 ton/m³, i.e., at the Southeastern Plant Environment Laboratories, although in practice less than 0.1 ton/m³ would be satisfactory at that latitude.

Greenhouse Lighting

Photoperiod-control lighting should be installed in all greenhouses and used to keep the plants on long-day conditions. Reflectorized incandescent-filament flood lamps provide the best overall source for photoperiod control. Interrupting the dark period for about 3 hr near the middle, from about 11:00 P.M. to 2:00 A.M., with an illuminance of 300 lux or more provides an efficient long-day effect on nearly all plant species that have any photoperiodic sensitivity. Sources like the Lucalox high-intensity-discharge (HID) lamp probably work well also for photoperiod control, but because of the large wattages more effort must be made to provide uniform illumination.

Greenhouse lighting to supplement the low natural light levels of winter has been an unresolved problem. Fluorescent lamps have largely proven unsatisfactory because an installation with enough lamps to provide a satisfactory light level also creates so much shade that natural light is severely restricted. The phytotron in Stockholm overcomes this problem by moving the lamps in and out of the greenhouse as a function of natural light intensity (Wettstein 1967).

The HID lamp is being used as the obvious solution for greenhouse supplemental lighting, and some success has been reported (Pinchbeck et al. 1971). In northern latitudes, natural light levels can be so low that almost any kind and amount of artificial light will improve plant growth. Some of our data indicate about 200 hlx would be necessary for optimum results. These levels can be reached with HID lamps and still allow room for a substantial amount of solar radiation to penetrate to plant level. Unfortunately the amount of power required is usually prohibitively expensive or simply unavailable. Recent efforts to reduce the power requirements (Smith, Downs, and Jividen 1973) have successfully used a moving HID lamp system to provide additional photosynthetic light during winter.

References

Alberda, L. 1958. The phytotron of the Institute for Biological and Chemical Research on Field Crops and Herbage at Wageningen. *Acta Bot. Neer.* 7:265.

ASHRAE Guide and Data Book: Fundamentals and Equipment. 1965. New York: American Society of Heating, Refrigerating, and Ventilating Engineers.

ASHRAE Handbook: 1973 Systems. 1973. New York: American Society of Heating, Refrigerating, and Ventilating Engineers.

Bailey, W. A. 1965. Fan and pad cooling of greenhouses. *Acta Hort.* 6:109–21.

Björkman, O., C. Florell, P. Nolmgren, and A. Nygren. 1959. The phytotron at the Institute of Genetics, Uppsala, Sweden. *Ann. Acad. Regal Sci. Upsal.* 3:5–20.

Bouillène, R., and M. Bouillène-Wolrand. 1950. Le Phytotron de l'Institut de Botanique de l'Univerisité de Liège, Belgique. *Arch. Inst. Bot.* 20:1–61.

Braak, J. P., and L. Smeets. 1956. The phytotron of the Institute of Horticultural Plant Breeding at Wageningen. *Euphytica* 5:205–17.

Bretschneider-Herrmann, B. 1962. Das Phytotron in Rauischholzhausen. *Acker Pflanzenbau* 115:213–22.

Busser, J. H. 1971. *Controlled Environment Enclosure Guidelines.* BIAC Information Module M21. Washington, D.C.: American Institute of Biological Science.

Canham, A. E. 1965. Automatic greenhouse shading. *Acta Hort.* 2:71–76.

Chouard, P., R. Jacques, and N. de Bildering. 1972. Phytotrons and phytotronics. *Endeavour* 31:41–45.

Climate Laboratory Operating Manual. 1973. Palmerston North, New Zealand: Department of Scientific and Industrial Research, Plant Physiology Division.

Doorenbos, J. 1964. The phytotron of the Laboratory of Horticulture, State Agricultural College, Wageningen. *Mededlingen Directeur van de Tuinbouw* 27:432–37.

Downs, R. J., and W. H. Bailey. 1967. Control of illumination for plant growth. In *Methods in Developmental Biology,* ed. F. H. Wilt and N. K. Wessels, pp. 635–45. New York: Thomas Y. Crowell.

Heck, W. W., J. A. Dunning, and H. Johnson. 1968. *Design of a Simple Plant Exposure Chamber.* National Center for Air Pollution Control publication APTD-68-6. Cincinnati, Ohio: U.S. Department of Health, Education and Welfare.

Hinkle, C. N., C. K. Spillman, and L. C. Shenberger. 1968. Maintaining 100% relative humidity in an alternating-temperature seed germination chamber. Paper no. 68-926, American Society of Agricultural Engineers, St. Joseph, Mich.

How to calculate greenhouse heat loss. 1962. *The Exchange Magazine for the Flower, Nursery and Garden Center Trade* July, pp. 1–6.

Hughes, A. P. 1964. Scientific aims of the phytotron of Reading. *Phytotronique* 1:7–9. Paris: Centre National de la Recherche Scientifique.

Issacs, G. W., L. C. Shenberger, and A. S. Stone. 1952. The design and development of an alternating temperature seed germinator. *Proc. Assoc. Off. Seed Anal.* 42:154–60.

King, G. R. 1971. *Modern Refrigeration Practices.* New York: McGraw-Hill Book Co.

Kramer, P. J., H. Hellmers, and R. J. Downs. 1970. SEPEL: New phytotrons for environmental research. *BioScience* 20:1201–4.

Lang, A. 1963. Phytotron design criteria: Biological principles. In *Engineering Aspects of Environment Control for Plant Growth,* pp. 5–20. Melbourne, Australia: CSIRO.

Lawrence, W. J. C. 1957. The glasshouse as an environment for plant experiments. In *Control of the Plant Environment,* ed. J. P. Hudson, pp. 129–38. London: Butterworth Scientific Publications.

Matsui, T. 1972. Biotron Institute, Kyushu University. In *Phytotrons and Growth Cabinets in Japan,* pp. 15–26. N.p., Japanese Society of Environmental Control in Biology.

Morris, L. G., and R. A. H. Johnson. 1965. Some factors in the design of metal structures. *Acta Hort.* 2:16–26.

Morse, R. N., and L. T. Evans. 1962. Design and development of CERES: An Australian phytotron. *J. Agr. Eng. Res.* 7:128–40.

Nelson, S. O., W. W. Wolf, and R. B. Stone. 1963. Seed germinating chamber control. *Trans. ASAE* 6:158–62.

Nisen, A. 1965. Construction, orientation et forme des series. *Acta Hort.* 2:7–15.

Pinchbeck, W., F. K. Johnson, D. N. Stiles, and S. J. Noesen. 1971. *Increased Production of Forever Yours Roses with Supplemental Lighting.* General Electric Technical Information Series no. 71-0L-001.

Rajki, S. 1973. Research strategy of the Martonvassar Phytotron (Hungary). *Phytotronic Newsletters* nos. 4, 5, 6, pp. 42–46, ed. P. Chouard and N. de Bilderling. Gif-sur-Yvette, France.

Seeman, J. 1952. Strahlungsverhaltnisse im Gewachhausen. *Arch. Meteorol. Geophys. Bioklimatol.* Ser. B, 4:193–206.

Senn, H. A., D. P. Anderson, and L. C. Anderson. 1965. *Manual for Investigators.* Madison: University of Wisconsin Biotron.

Smith, W. T., R. J. Downs, and G. M. Jividen. 1973. Economical HID source for greenhouse light supplement. ASHS abstract. 70th meeting, Raleigh, N.C.

Standards for ventilating and cooling greenhouses. 1962. *The Exchange Magazine for the Flower, Nursery and Garden Center Trade* July, pp. 1–6.

Stocker, O. 1949. Grundlagen einer naturgemassen Gewachshauskultur. *Grundlagen u Fortschritte i Gartenbau.* Stuttgart: Verlag Ulmer.

Trane Air Conditioning Manual. 1968. LaCrosse, Wis.: Trane Co.

Tumanov, I. I. 1959a. The Soviet phytotron. *Priroda* 1:112–17.

Tumanov, I. J. 1959b. The first year of functioning of the Soviet phytotron (in Russian). *Akad. Nauk. SSSR Izv.* Ser. Biol., 1959:265–82.

Verhey, I. C. 1955. Germination equipment of the seed testing station at Wageningen. *Proc. Internat. Seed Test. Assoc.* 20:5–28.

Walker, J. N., and J. W. Abernathie. 1964. Evaporative cooling of greenhouses. *Amer. Orchid Soc. Bull.* 33:377–81.

Went, F. W. 1957. *Environmental Control of Plant Growth.* Chronica Botanica, vol. 17. New York: Ronald Press.

Wettstein, D. von. 1967. The phytotron in Stockholm. *Stud. Forest. Suec.* 44:1–23.

Whittle, R. M., and W. J. C. Lawrence. 1959. The climatology of glasshouses. I. Natural illumination. *J. Agr. Eng. Res.* 4/326–40.

Yamamoto, T. 1972. The Hokkaido National Agricultural Experiment Station Phytotron. *Phytotronique* 2:305–16. Paris: Gauthier-Villars.

2 CONDITIONING SYSTEMS FOR THE MAJOR ENVIRONMENTAL PARAMETERS

Temperature Control

The performance of a controlled-environment facility is, of course, a function of the total system design. Assuming that the various components are sized correctly, the temperature control devices are perhaps the most important part of that total system design. Not only does temperature profoundly influence plant growth and development, but also it affects the degree of control of other environmental factors such as relative humidity.

Temperature in commercial controlled-environment rooms varies about the set point from ± 0.2 to ± 1.7 C, depending on the manufacturer. The temperature differentials reported by manufacturer do not necessarily reflect the degree of control that might be measured by the owner, because the system of measurement is seldom reported. Thus a slow-response instrument such as a thermograph may indicate ± 0.2 C control whereas a faster-response instrument will show ± 2.7 C variation in the same location. Methods of environmental measurement will be discussed in chapter 3.

Generally, the lower the radiant energy and the higher the air flow, the more precisely can temperature be controlled. In a plant-growth chamber, however, the designer is faced with very high radiant energies, and the allowable air velocity is usually limited to about 30 m/min. Nevertheless, control of the order of ± 0.2 C about the set point can be readily attained.

Coil Temperature

A steady cooling-coil temperature that approaches the room or space temperature will result in the least variation in room ambient conditions. Once the desired air temperature has been reached, the steady load conditions usually found in the controlled-environment room should not require any sudden or large changes in coil temperature. Clearly, keeping the coil-to-room temperature differential relatively small and steady requires some kind of modulated control. An on-off system fails to work satisfactorily because when the coil is flooded with cold coolant or heating devices are activated, the large residual effects cause under- or overshooting of the set point.

Secondary-coolant systems usually use three-way proportioning valves that mix warm return coolant "A" with the cold supply "B" in whatever ratio "C" will establish a coil temperature just sufficient to handle the heat loads. A point sometimes not understood is that satisfactory operation of this kind of system requires equal pressure on the A and B sides of the valve. Unequal pressure can result in the valve "hunting," or continuously adjusting the A and B sides in an effort to reach a balance point, and temperature control will vary more than it should.

Direct-expansion units resort to some kind of compressor capacity control to prevent on-off operation. Cylinder unloading may be practiced on large compressors, but maximum cylinder unloading is seldom adequate in growth-chamber applications. Unloading is a practice whereby a drop in suction pressure actuates a device that will hold the suction valves open. Thus vapor pulled into the cylinder on the suction stroke is forced back through the open valves on the compression stroke. The most popular method of capacity control is hot gas bypass, in which the high-pressure gas from the compressor discharge is bypassed through a reverse flow valve to the compressor suction line. This method maintains a relatively constant suction pressure so that the compressor runs continuously. Additional control is sometimes obtained by liquid injection; low-pressure liquid is bypassed from the evaporator inlet to the suction side of the compressor to desuperheat the suction gas, thereby preventing excessively hot cylinder heads.

Control of direct-expansion cooling-coil temperature can be improved by bypassing the hot gas from the discharge of the compressor to the evaporator or cooling-coil inlet (fig. 1.7). This piping system acts like an artificial evaporator load. The expansion valve reacts to the artificial load and maintains superheat control at the evaporator outlet. Although these systems prevent compressor cycling, the control method still technically has two stages. However, adjusting the ratio of liquid to hot gas as a function of heat load can result in a steady cooling-coil temperature very near the set point of the room. A recent modification of this system is called "two-phase direct expansion" and replaces the expansion valve and hot gas bypass valve with a single, multipoint, modulating valve.

Controllers

Setting up an adequate refrigeration system that has the ability to maintain a proper coil temperature–load relationship still will not provide an acceptable degree of temperature uniformity unless the controls are properly selected and correctly installed. Temperature control depends directly on how well the temperature and the magnitude of the temperature changes are sensed and measured. Since temperature is a function of molecular movement due to the absorption or loss of heat energy, matter expands or contracts in volume, electrical resistivity changes, fluids exert differing pressures, and viscosity of liquids changes. Temperature can be measured and controlled by evaluating any of these physical changes, and a survey of existing and currently manufactured plant-growth chambers shows a wide array of temperature sensors in use.

The bimetal switch. The most common type of temperature controller is the bimetal switch (fig. 2.1). Two thin strips of dissimilar metals are fastened together. One metal has a high coefficient of expansion (e.g., brass $= 315 \times 10^{-7}$) and the other a low coefficient (e.g., In-

FIG. 2.1. Bimetal thermostats operating, at left, adjustable electrical contacts (Barber Colman Co., Rockford, Ill.) and, at right, mercury switches (Honeywell, Minneapolis, Minn.).

var $= 15 \times 10^{-7}$). Brass therefore expands about 20 times as much as Invar when the temperature increases. To equalize internal stresses the bimetal strip will bend in the direction of the metal with a low expansion coefficient, the Invar, and the degree of bending will be proportional to the temperature change. Adjustable electrical contacts allow the movement of the bimetal strip to open or close an electrical circuit for heating or cooling. The bimetal thermostat is inexpensive and reliable and has been used successfully in older controlled-environment rooms with on-off control, as well as in seed germinators, incubators, and drying ovens. Response speed is slow even when the switch is placed in an aspirated enclosure, and temperature variation of several degrees can occur in large rooms. The bimetal switch has a high degree of reliability, however, and will continue to be used in many applications.

Gas-filled bellows. Somewhat better temperature control can be obtained with a bellows type of thermostat. In this device an increase in temperature causes expansion of a gas inside a metallic bellows. The resultant movement of the bellows can operate an arm or slide wire of a variable resistor. The resistance value then determines the position of a proportioning valve by means of an electrical controller. The resulting proportional control provides much better temperature maintenance than on-off systems, and several satisfactory growth rooms use this control and sensor method. The sensor also has a slow response time, and a slow-speed valve must be used to prevent hunting from full open to full close, which results in on-off operation and unsatisfactory control and defeats the purpose of the proportioning valve. Reliability is very good, and response rate can be improved by aspiration.

Hydraulic thermostats. Hydraulic thermostats are pressure-actuated systems that use the thermometer principle—the change in level of a liquid or gas in a fine-bore tube as a function of temperature change (fig. 2.2). The pressure exerted by the gas or liquid operates a mechanical linkage to a controlling device such as a set of limit switches or a variable resistance. The finer the bulb bore, the larger the ratio of surface area to volume and the faster the response. Bulbs filled with an organic liquid such as alcohol or ethyl benzene are sometimes used in plant-growth rooms. Response times are slow, usually about 1 min even with long, coiled bulbs in an air stream. Instruments using gas-filled bulbs have only

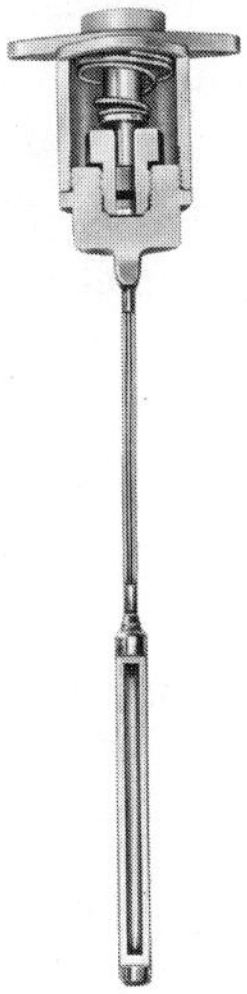

FIG. 2.2 Mercury-filled thermal sensing element. Capillary can be any length up to 36.5 m (Partlow Corp., New Hartford, N.Y.).

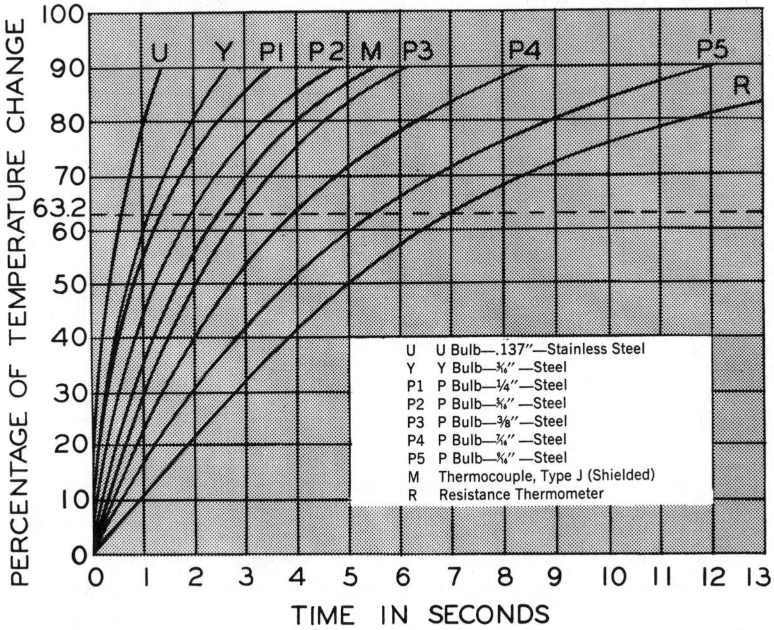
PERCENTAGE OF TEMPERATURE CHANGE
100
90
80
70
63.2
60
50
40
30
20
10
0
0 1 2 3 4 5 6 7 8 9 10 11 12 13
TIME IN SECONDS
U Y P1 P2 M P3 P4 P5 R
U U Bulb—.137″—Stainless Steel
Y Y Bulb—3/16″—Steel
P1 P Bulb—1/4″—Steel
P2 P Bulb—5/16″—Steel
P3 P Bulb—3/8″—Steel
P4 P Bulb—7/16″—Steel
P5 P Bulb—9/16″—Steel
M Thermocouple, Type J (Shielded)
R Resistance Thermometer

FIG. 2.3. Response rates of various thermal sensing element configurations (Partlow Corp., New Hartford, N.Y.).

slightly better response. Mercury-filled units seem to provide the best response of this type of sensor. Tubular bulbs are usually about 0.9×66 cm, with a response time of 4–5 sec. Coiled capillary bulbs 0.5×368 cm are reported to have a 1-sec response (fig. 2.3). Our own experience with this type of system indicates that in growth-chamber applications the response is not that fast. Drift can be a problem, so the units should be recalibrated frequently. Although we have had some technical problems with these kinds of sensors, they are basically satisfactory. Other investigators using various makes and models have also reported satisfactory reliability. The hydraulic system is used on many controlled-environment rooms and is the device that feeds temperature information into many of the precut-cam type programmers.

Resistive temperature transducers. In general the electrical resistance of metals increases with rising temperature. The resistance variation as measured with a Wheatstone bridge arrangement is a direct indication of the temperature. The bridge senses a resistance change, which is translated into a voltage output. The voltage output drives a valve actuator that includes a potentiometer used as a rebalancing circuit (fig. 2.4).

Resistance thermomenters are the most accurate of the temperature transducers used in controlled-environment applications. Response time is very rapid. For air measurements, phosphor-bronze alloy wire is wound on an open or "belt-buckle" bobbin so that air can pass freely around almost the entire wire length (fig. 2.5). Such resistance elements are relatively large, however, and extra effort is required to obtain satisfactory shielding from radiant energy.

In plant-growth-chamber applications we have found several disadvantages to resistance-element systems. High humidity can cause leakage resistance, and the least bit of corrosion at any of the electrical connections produces a sizable error in temperature control. Dirt, lint, and bits of peat moss accumulate on the wire and produce a significant decrease in responsiveness. For unexplained reasons, when the resistance wire is wound on an open bobbin and placed in an air stream the wire has a marked tendency to break as it ages, with, of course, complete loss of temperature control. Many of the commercial units are d-c bridges and amplifiers which can, and do, pick up a-c noise components from fans, fluorescent lamps, and other equipment, creating false signals and in-

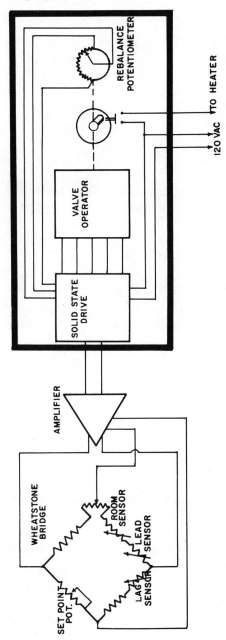

FIG. 2.4. Schematic drawing of a temperature-control system using lead-lag resistance elements.

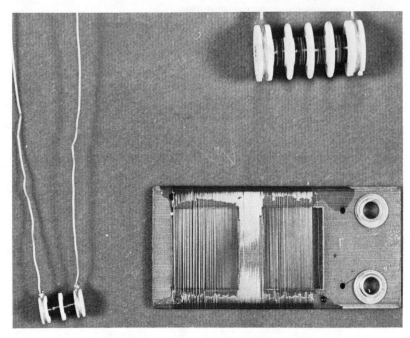

FIG. 2.5. Three types of resistance-element temperature transducers.

creasing the variation about the set point. Slight changes in the total resistance of the rebalancing circuit, which is usually placed at a distance
from the sensor and bridge, also create control problems and in fact are
among the main trouble points of the system. Rebalancing slide wires in
some commercial controls have to be replaced every 6 months.

Semiconducting-resistance transducers. Usually called thermistors, these semiconductors have a large negative resistance-temperature
coefficient of several percent change in resistance per unit of temperature
change. The time constant of commercial thermistors is relatively fast,
they are physically rugged, and because of their low specific heat they
draw practically no heat from the material being measured. Small size
contributes to more efficient shielding from radiant energy and to smaller
aspirators. Thermistors have a high impedence that eliminates many problems of circuit design and in large measure eliminates the influence of
lead-wire resistance and small changes in resistance due to corrosion.

Thermoelectric transducers. If two dissimilar metals are joined together, an electromotive force (emf) arises whenever two junction points, P_1 and P_2, are at different temperatures, t_1 and t_2. The thermal emf is essentially a linear function of the temperature difference between the two junctions and can therefore be used to measure the temperature difference, $t_1 - t_2$, between P_1 and P_2. The thermal emf is usually measured with a potentiometer, which may be calibrated in terms of temperature degrees; the emf can also be translated to temperature terms from tables.

The temperature at the measuring point can only be determined if the temperature at the reference junction is known and constant. For growth-chamber applications, reference-point temperature compensation is usually done electronically. One method is to place the reference junction next to a temperature-sensitive resistor or thermistor which forms one leg of a balanced bridge. If the bridge becomes unbalanced because of a change in reference-junction temperature, an output voltage results to compensate for the change in thermal emf of the junction.

Thermocouples have a fast response, can be very small, and seem to have few maintenance problems. The newer solid-state controlling potentiometers are compact systems with good reliability.

Location of Sensors

Greater sensitivity usually results when the sensor is placed in an air stream. In controlled-environment-room applications the sensor may be in the air return duct or in an aspirated housing near the plant growing area. Some effort should be made to assure that air flows are adequate and that the element is completely shielded from radiant energy. Obviously, the air should be pulled instead of pushed past the sensor to avoid heat generated by the fan motor. No data seem to be available on the air-flow requirements, but aspirators that have flows of 45–60 m/min seem to result in good sensor sensitivity. All bodies at all temperatures are continuously radiating energy to each other. Those at constant temperature radiate as much as they receive. The amount of radiant heat received per unit of time, E_{Tot}, is the sum of the energy absorbed, E_a, reflected, E_r, and transmitted, E_t. Therefore, a good reflector is a poor absorber, as is a substance with high transmittance values.

Design of the aspirator housing must therefore consider the absorption of heat and radiant energy by the housing, with subsequent convection and conduction to the sensor. Fuchs and Tanner (1965) showed that although mirrors provided the best surfaces for aspirator housings, aluminized mylar and highly reflective white paint gave very satisfactory results. The best commercial aspirated systems use a double-wall housing with the outer wall painted with a high reflectance epoxy enamel.

Because of the volume of the controlled-environment room and the relatively slow rate of air exchange, an appreciable time elapses between the sensing of a load change, chilling or heating of the air to compensate for the change, distribution of the air throughout the room, and finally sensing that the demand has been met. During this elapsed time period, which may be of the order of 10–15 sec, the sensor keeps demanding an adjustment and the valve operator continues to open or close the valve. The result is too much or too little coolant in the coil, with a consequent overshoot of the control point and a greater than desired temperature fluctuation. An additional sensor placed in the discharge air will anticipate the change for the room sensor and prevent continued demand. Many commercial controllers using resistance elements or thermistors are set up to use lead-lag sensors, and some have devices that allow the percentage of authority between room or main sensor and the lead-lag units to be adjusted experimentally until the correct ratio is obtained for each specific facility or design (fig. 2.4). Commercial components for proportional control with dual sensors can result in ±0.25 C control. Since these components are usually solid-state devices, the amplifiers, bridges, and sensors are easily exchangeable if they malfunction. Some planning is, of course, required to ensure rapid accessibility and ease of calibration and balancing. Use of anticipating sensors also works well to control temperature in greenhouses (fig. 2.6).

Thermocouple controllers do not use two sensors but have an approach feature that modifies the reset action and prevents overshoot when a time lag exists before changes can be noted by the sensor. Tests of this type of controller have been so favorable that thermocouple systems will probably be used much more widely in the future for growth-chamber temperature control.

Irrespective of the type of system used, commerical division components are nearly always a poor choice. Commercial divisions seem especially prone to altering components or suddenly stopping production entirely. A recent modification by the manufacturer to the valve operator

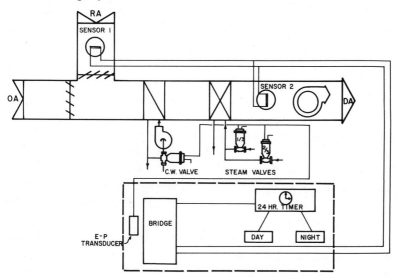

FIG. 2.6. Schematic drawing of air-handling system for an air-conditioned greenhouse, showing sensor location (*OA* = outside air, *RA* = return air, and *DA* = discharge air; *E-P* = electropneumatic).

controller shown in figure 2.4, presumably for improving control of building air conditioning, resulted in such poor growth-chamber performance that the entire system had to be replaced. Many of the commercial division components, like the valve controller, cannot be repaired because of encapsulation or because wiring diagrams are not available. Industrial division components may cost a little more initially, but when used in controlled-environment applications they are generally more reliable than commercial controls, require less maintenance, and can be rebuilt and repaired when necessary.

Relative Humidity

Many environmental physiologists believe that relative humidity control is of little importance in most plant-growth-room applications as long as the RH is kept above some minimum level such as 50% (Went 1957). Relative humidity levels between 60 and 90% certainly seem to exert only a small influence on plant growth and development. Investigations that deal with evaporation or transpiration require close humidity control, however, and some degree of control is frequently required in many other studies on such topics as translocation, air pollution, and insect behavior. Therefore, relative humidity control is in enough demand to warrant a discussion of the problems and methods of obtaining it.

Statements such as "the lower the water vapor content or relative humidity of the air the more rapidly will water be transpired" leave the impression that water vapor content and RH are synonymous and autonomous. As shown in table 2.1, a constant vapor pressure at different temperatures results in a wide range of RH values. Conversely, at a constant RH the vapor pressure and water vapor content vary rather dramatically as the temperature changes (table 2.2). Therefore, attempting to maintain constant RH conditions over a wide range of temperatures in-

TABLE 2.1 Effect of temperature on the relative humidity of air maintained at a constant vapor pressure of 12.70 mm Hg

Temperature (C)	RH (%)
15	100
20	72
25	53
30	39
35	30

TABLE 2.2 Effect of temperature on water content and vapor pressure of air maintained at constant 70% relative humidity

Temperature (C)	Water (g/kg dry air)	Vapor pressure (mm Hg)
15	7.4	8.95
20	10.3	12.28
25	14.0	16.63
30	18.8	22.27
35	25.5	29.53

troduces considerable variation in the vapor pressure and vapor pressure deficit. Identical RH values do not mean the same moisture conditions unless the temperature is constant. At different temperatures the only way to keep the same moisture conditions in the air is to have different relative humidities.

Anderson (1936) tried to promote the use of vapor pressure deficit as the only meaningful term since it influences evaporation in the same way irrespective of temperature. The vapor pressure deficit is the difference between the actual amounts of water vapor inside and outside the leaf and consequently provides an indication of the gradient involved. As the temperature increases the vapor pressure deficit increases if the relative humidity does not change. If the temperature of the leaf is higher than surrounding air, the gradient from leaf to air will become steeper. It is therefore possible to have transpiration occur when relative humidity of the air is at saturation. Most biologists certainly appreciate this argument and are interested in relative humidity only as a means of regulating the vapor pressure gradient from leaf to air.

Nevertheless, in the phytotron we still get requests for constant relative humidity over a wide range of temperatures. It becomes immediately apparent from examination of the psychrometric chart that to maintain an 80% RH at 30 C would require 250% more water in the air than 80% at 15 C. It is equally apparent that attempts to maintain a constant vapor pressure such as 12.28 mm Hg, which represents 70% RH at 20 C, would require over 100% RH at 14 C and only 38% at 30 C. A constant gradient between leaf and atmosphere as represented by vapor pressure deficit would, however, be well within the realm of practicality between 15 and 30 C (fig. 2.7).

Humidification

Control of relative humidity requires a humidifier system, a dehumidifier, and a proper set of transducers and controllers. When installed properly, mist nozzles, wet pads or hot water evaporators will provide adequate humidification. Steam injection is the most efficient humidification system, however, provided the steam is clean. Steam supplied from central boiler plants usually contains additives to prevent corrosion. When such steam is injected into the growth-chamber air stream, some types of additives will condense as a talc-like deposit on the plant

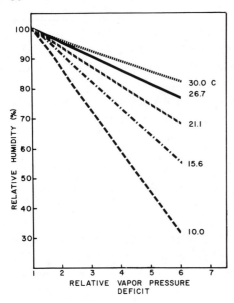

FIG. 2.7 The relationship of vapor-pressure deficit to relative humidity and temperature. Data were obtained from an ordinary psychrometric chart.

leaves. There is, of course, a distinct possibility that the corrosion-preventing additives may be detrimental to plant growth and development. Treated steam should therefore be avoided, and small steam generators, available from several commercial sources, should be used instead. Steam injection is an isothermal action that induces the least disturbance of temperature conditions. Response is fast, and injection rate can be controlled in a proportioning manner.

Water atomization is a diabatic process and will tend to cool the air. If mist nozzles or wet pads operate on the discharge side of the cooling coil, the evaporative cooling effect can introduce a temperature variation of 4–6 C. If they are placed on the inlet side of the cooling coil, the degree of humidification becomes an immediate function of the coil temperature. If a coil temperature of 23 C is required to maintain 27 C in a lighted growth room, the highest relative humidity that can be obtained is about 82%. When the chamber is dark, the coil-to-room temperature differential is usually less, and 95% RH can be obtained. The same room operated to give an air temperature of 10 C may require a 4-C coil, and the maximum RH attainable is only 66%. The lower coil temperature is required in part because of the warmer spray water. At temperatures higher than 35 C, the coil may be only 33 C but 95% RH may not be achievable because the water temperature is probably lower than that of the coil.

Since evaporation increases with temperature in about the same ratio as does maximum vapor pressure, it seems clear that warmer water will humidify better than cold water. Less obvious, but equally well known, is the fact that spray nozzles act as dehumidifiers when the water temperature is lower than the dew point of the room air. For example, if the dry-bulb temperature of the room air is 30 C and the wet-bulb temperature (that at which evaporation of water can bring the air to saturation) is 25 C, a spray-water temperature of less than 24 C will dehumidify.

A water spray system using mechanical slingers or centrifugal atomizers produces much larger water droplets than do air-driven mist nozzles. Much of the water is not absorbed by the air; as it recirculates it acts like a heating or cooling coil and can produce a large temperature disturbance because a large amount of heat is transferred to the air as sensible heat. Capacity of the system is usually low, and response is slow.

Dehumidification

Close control of RH at various set points requires a dehumidification system to oppose the humidifier, plant evapotranspiration, and the spillage that occurs during the daily watering process. A popular method has been to use the cooling coil as a dehumidifier, with electrical reheaters to maintain temperature. Examination of a room designed in this manner showed that the system worked rather well when the room temperature was between about 15 and 27 C. When the ambient temperature was reduced to 10 C, high humidity caused ice formation on the coil, with subsequent loss of temperature control. This is hardly surprising, since the psychrometric chart clearly shows that the coil temperature must be below freezing to dehumidify the air below 50% RH at 10 C. Above 27 C, with an RH of less than 60%, the cooling effect of the evaporator during dehumidification was greater than the capacity of the heaters, and the high-temperature set points could not be obtained.

Effective humidity control can be obtained by using a cold plate or cold coil with a large surface area as a dehumidifier in conjunction with a separate evaporator for temperature control. The cold plate can be used to dehumidify the air to a selected dew point, after which the air is reheated to the desired dry-bulb set point. Sometimes a bypass system is used to provide modulated control by passing a calculated percentage of the room air or outside makeup air through the humidity control system while the remainder circulates through the temperature control apparatus. The cold plate cannot be operated much lower than 5 C, which means that at an air temperature of 10 C the lowest humidity possible would be only about 45%.

Theoretically the chemical drier should be a satisfactory method of dehumidification and may be the only practical method at temperatures below freezing. Either two driers or a dual-bed, regenerating drier must be used so that one bed or drier can be regenerated while the other is dehumidifying. The main disadvantage of chemical driers is the large size needed to remove the large water loads that accumulate in the plant-growth chamber. One commerical chamber manufacturer has kept drier size to economical proportions by adding a precooling coil. Operation of the precooling coil and removal of heat generated by the chemical action of drying require a larger refrigeration system than would be necessary on a growth chamber without such a dehumidification system.

Controllers

As is true with most environmental factors, our ability to control relative humidity is largely a function of our ability to measure or sense it. The most common hygrometer is made from human hairs linked by a series of levers to a switch or set of electrical contacts (fig. 2.8). The hairs must be kept clean and intact. The hair humidistat has low sensitivity, considerable lag, and precision of at most 6%. Although the unit requires frequent recalibration, reliability is generally good. Sensitivity, while still poor, can be greatly improved by mounting the humidistat in an aspirated cabinet.

Resistance-variation devices for humidity sensing consist essentially of a hygroscopic body that absorbs water from the air and a salt that dissociates in the presence of water. Wires are embedded in the hygroscopic body, and the concentration of the dissociated salts (which is proportional to the relative humidity) is determined by measuring the electrical resistance of the element (fig. 2.9). A number of hygroscopic materials such as lithium, calcium, and zinc chlorides have been used in these devices. The measuring current has to be kept small to avoid heating the probe, and some kind of temperature compensation is required

FIG. 2.8. Human hair humidistat operating a snap action switch to control a humidification system (Honeywell, Minneapolis, Minn.).

FIG. 2.9. Lithium-chloride type of humidity sensor and solid-state bridge and amplifier (Barber Colman Co., Rockford, Ill.).

since both water absorption and dissociation vary with temperature. The response speed of resistance units is relatively high, and commercial ones are available with an error of less than 1.5%. Resistance-variation sensors can operate on-off controllers that actuate relays to humidifiers and dehumidifiers, or they can operate proportional systems. The main problem is the failure of the sensor whenever liquid water condenses on it, a frequent occurrence in many growth-chamber applications where the temperature is lower at night than during the day and where the change is made rapidly.

Humidity control by thermal systems, sometimes called Dunmore hygrometers or Dewcels (a trademark of the Foxboro Co., Foxboro, Mass.) permits an absolute rather than a relative measurement by determining the dew-point temperature. Essentially the Dewcel is a tubular wick impregnated with lithium chloride and mounted over a metal tube that is electrically insulated from the wick. Two parallel wires are wound on the wick and connected to an a-c source. The lithium chloride absorbs water from the air and becomes conductive, allowing a current to pass between the two wires through the lithium chloride layer. The current generates heat, which tends to evaporate the water and consequently leads to a reduction of conductivity and current. At equilibrium the temperature of the lithium chloride produces a partial pressure of water over a saturated lithium chloride solution that just equals the ambient vapor pressure. The temperature is measured by a thermistor or resistance element. Since the vapor pressure–temperature relationship of lithium chloride is accurately known, the output can be calibrated directly in dew-point temperatures. The signal from the transducer can be used to operate on-off or proportional controllers. The principal disadvantage again is the failure of the sensor when liquid water condenses on it.

The control of relative humidity remains a compromise between cost and experimental value. Very little dehumidification occurs in chambers that have a modulating control that maintains evaporator temperature just a few degrees below the set point for the space. In rooms with such systems, and plant-watering methods that result in considerable spillage and water draining through pots onto the floor, the RH rarely falls below 70%. Where drainage and spillage are not factors and where the cooling systems use large-area, cold coils, a simple humidification system may be needed to keep the RH at a satisfactory 70% level. Where precise control of humidity is required, costs begin to rise dramatically. Consider that at moderate temperatures every 0.5-C variation in temperature results in a 3% shift in RH without any change in moisture content. If a room temperature varies ± 0.5 C, then humidity will automatically vary ± 3%. Moreover, the difficulty of maintaining constant RH when the night and day temperatures vary may require very large dehumidifier systems. For example, a room operating at 65% RH during a 27-C day will show 90% RH if the night temperature drops to 21 C, although there is actually no change in water content or vapor pressure. To maintain a constant RH of 65%, the dehumidifier must decrease the vapor pressure by 30%. Consider also the large amounts of water that must be removed if the usual "slop-culture" watering methods are used. Two 0.61 ×

4.88-m coils used in the Power Groove room at Beltsville, Maryland, removed over 12 l. of water from the air after every plant watering, but nonetheless a several percent increase in RH occurred.

Light

How much light is required in a plant-growth chamber and what are the best light sources for plant growth? These are questions frequently asked by persons contemplating the purchase or construction of plant-growth facilities.

Intensity

Solar radiation reaches levels of about 1.35 cal cm^{-2} min^{-1}, which provides an illumination of about 1100 hlx in the afternoon on a clear summer day in temperate zones. These light levels are only maintained for a few hours each day and are not reached on all days. One could argue that the total energy per day would have more biological significance, and certainly the frequency of distribution of various light levels during the growing season is important ecologically (Gaastra 1964). Table 2.3 shows the relation of between growth-chamber energy and natural light. Although peak natural intensities do not seem as biologically important as other values of photosynthetically active radiation (PAR), there nevertheless appears to be a marked tendency among biologists to use such numbers as desirable light intensities for growth chambers. The ideal growth room would perhaps be able to reach such light levels, but both initial and operating costs soon raise the question of how much light the plants really use efficiently. Saturating light intensities for dry-weight production are reported to be 160 hlx for tomato plants (Went 1957) and 270 hlx for wheat (Friend, Fischer, and Helson 1963). The rate of photosynthesis under normal CO_2 conditions also fails to increase appreciably after light levels reach about 270 hlx (Bohning and Burnside 1956, Blackman and Black 1959, Gaastra 1959).

Light-saturation data are often taken using a single leaf or from plants carefully lighted to avoid self-shading. Under most growing conditions some degree of self-shading occurs, and the saturation light levels increase considerably (Thomas and Hill 1937, 1949, Müller 1951, Went

TABLE 2.3 Comparison of growth chamber and solar photo-synthetically active radiation (PAR) as percentage of days on which the natural light levels exceed those in growth chambers

| | Artificial light source | | | |
| | TL 120[a] | | HPL 400[b] | |
Month	Energy/day	Peak energy	Energy/day	Peak energy
January	0	0	0	0
February	0	16	0	3
March	0	48	0	17
April	13	70	0	50
June	48	75	5	55
July	40	—	2	—
August	20	70	0	50
September	0	58	0	34
October	0	40	0	10
November	0	5	0	0
December	0	0	0	0

[a] Irradiance of 13.2×10^4 ergs cm^{-2} sec^{-1}, Philips' fluorescent lamps.
[b] Irradiance of 20×10^4 ergs cm^2 sec^{-1}, phosphor-coated mercury lamps.
Note: Adapted from Gaastra (1964).

1957). Our own data indicate that 215 hlx are about the minimum for good plant growth in controlled-environment rooms, and that 430–540 hlx will produce a high rate of growth, with the resulting plant closely resembling those grown under natural conditions.

Spectral Distribution

High levels of illuminance have been available for years in the form of mercury vapor lamps. Yet they have rarely been used in plant-growth-chamber applications because the spectral distribution resulted in unsatisfactory plant development. Improvements in high-intensity-discharge (HID) lamps, however, have resulted in the color-improved mercury, metal-halide (Multi Vapor, Metal Arc), and high-pressure sodium lamps (Lucalox, Cermalux). Growth chambers have been built with every type. The spectral distribution from a 1:1 mix of the light from Lucalox and mercury lamps has provided plant development at least as satisfactory as that obtained from fluorescent-incandescent systems (Jividen, Downs, and Smith 1970). Few comparative data are available, however, on the relative effectiveness of the individual types of HID lamps.

Meijer (1971) reports that on the basis of equal installed watts the high-pressure sodium lamp produces more plant weight than do other kinds of HID or fluorescent lamps. Additional research (Downs, Smith, and Jividen 1973) confirms these observations but shows that at light levels normally encountered in modern plant-growth rooms (i.e., 430–480 hlx) the sodium HID lamp induces excessive elongation whereas plants grown under metal halide lamps are shorter than would be considered normal. The most normal plants were obtained with light from metal halide lamps plus about the same amount of incandescent light as in the fluorescent-incandescent system.

Growth chambers have also been constructed using xenon arc lamps (Senn, Anderson, and Anderson 1965, Plapper 1972), usually because of the flat spectral distribution throughout the visible portion of the spectrum (fig. 2.10). Considerable care must be taken with these lamps, however, because they form large amounts of ozone and they operate at temperatures hot enough to melt acrylic plastic barriers.

Color temperature is rarely considered in selection of lamps for plant growth, although it appears to be an excellent predictor of the way a particular light source might perform when used for that purpose. The basis for determining color temperature is visual comparison between the light source and a black body radiating at various temperatures. Although this method works satisfactorily with an incandescent source, the spectral irradiance of fluorescent lamps is so different from that of a black body that only a correlated color temperature can be obtained. Nevertheless the data are adequate and the position of the lamps on the color temperature scale can be shown (fig. 2.11).

The most desirable spectral distribution would probably be similar to that of sunlight (fig. 2.12), if only because all the wavelengths are present at a sufficient level to operate all the photochemical systems of the plant. Most plant-growth chambers are lighted, however, with a combination of fluorescent and incandescent lamps that does not approximate a natural spectral distribution (fig. 2.13). The incandescent lamps usually provide about 10% of the total illuminance. A rule-of-thumb method has been to install incandescent watts in an amount equal to about 30% of the

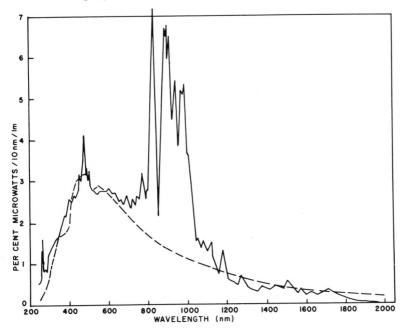

FIG. 2.10. Spectral energy distribution of a xenon arc lamp (solid line) compared to idealized solar spectrum (dashed) (drawing no. 480-LO-422, General Electric Co.).

fluorescent watts. Thus a small cabinet with 1600 w of fluorescent lamps would add 480 w of incandescent lamps in whatever number and size were necessary for uniform distribution. Whether this method results in the "desired" 10% incandescent illuminance depends on the efficiency of the fluorescent lamp being used as well as the rated and supply voltage for the incandescent source. The 1:10 ratio of incandescent to fluorescent light is usually attributed to early research by Parker (1946), Withrow and Withrow (1947), Parker and Borthwick (1949), and Went (1957). In fact, the optimum ratio was never investigated. H. A. Borthwick (personal communication, 1972) says that adding 10% of the illuminance as incandescent light increased dry-weight production as much as 100%, so this ratio was used and others were not tested.

The exact contribution of the incandescent light to plant growth

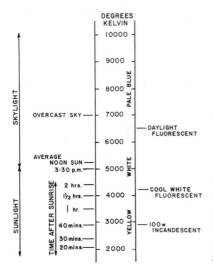

FIG. 2.11. Color temperature scale showing positions of some artificial light sources as compared to sunlight (General Electric Co., *Bull.* LD-1).

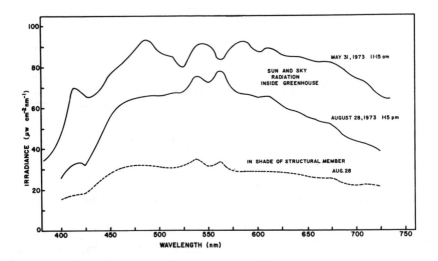

FIG. 2.12. Spectral energy distribution of sunlight and skylight inside an unshaded greenhouse. Measured with a spectroradiometer from Instrumentation Specialties Co. (ISCO) (Lincoln, Nebr.).

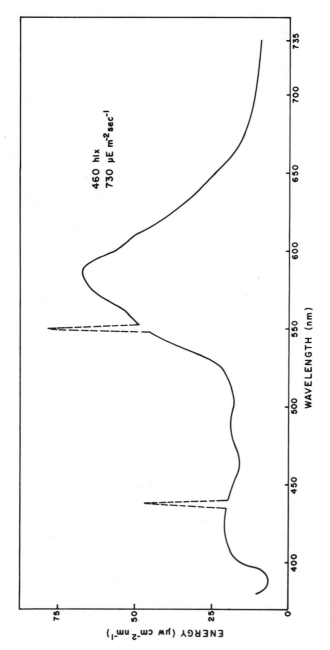

FIG. 2.13. Spectral energy distribution in a typical plant-growth room lighted by cool white fluorescent and incandescent lamps. Measured with an ISCO spectroradiometer.

is unknown. A part of the incandescent effect lies in an increase in soil temperature (Bailey and Downs, unpublished data). Plant temperature may also be increased. Another part of the effect would be to change the red–far red and blue–far red ratios (fig. 2.13), thereby affecting the photomorphogenic pigment systems.

Many kinds of fluorescent lamps are available for use in plant growth chambers and produce different spectral energy distributions (fig. 2.14). After testing many kinds of fluorescent lamps, Went (1957) concluded, "We can say that no better combination of commercially available lamps for growing plants in artificial light was found than a mixture of fluorescent (warm white) and incandescent light." The same results (unpublished) were found by H. A. Borthwick at Beltsville; tests at the North Carolina State University (NCSU) phytotron are in agreement, except that cool white lamps seemed to produce about the same plant growth as did warm white. On an equal-installed-watt basis, other kinds of fluorescent lamps such as daylight or even deluxe forms of cool and warm white are not as effective in producing plant growth as are the cool white lamps.

Lamps designed especially for plant growth have met with mixed success. Some experiments indicate that such lamps are superior to cool white ones in producing growth, while other experiments consistently indicate that less plant growth is obtained. Efforts at producing a superior plant-growth lamp are continuing, and some improvements are being made.

The physical characteristics of fluorescent lamps create problems for the user of the plant-growth chamber. The fluorescent lamp is a glass tube coated with a mixture of phosphors that fluoresce when exposed to ultraviolet radiation, plus additives called "activators." After the addition of a drop of mercury and an inert filling gas such as argon, cathodes coated with an emitter of barium, calcium, or strontium are placed at each end of the tube. When the lamp is operated, the cathodes are heated to induce the release of electrons. Actually, of course, the electrode acts as a cathode only while emitting electrons, on half of the a-c cycle; it acts as an anode on the other half of the cycle. The excess electrons ionize the starting or filling gas, reducing tube resistance so that an arc can be struck by applying a relatively low voltage difference between the cathodes. The arc current flows through mercury vapor, changing the energy levels of

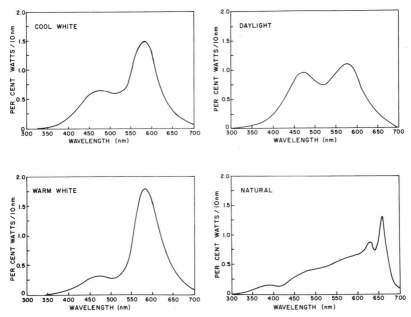

FIG. 2.14. Spectral energy distribution of four kinds of fluorescent lamps (GTE Sylvania, *Sylvania Eng. Bull.* 0-283).

the electrons in the mercury ions and releasing energy. Because the lamp is operated at a low pressure, the wavelength most efficiently generated by the arc is at 254 nm. The 254-nm radiation is absorbed by the phosphors, causing them to fluoresce and emit light. The phosphors have been carefully selected to respond most efficiently to ultraviolet radiation at the 254-nm maximum, and they do not operate effectively at other wavelengths.

The proportion of energy from the mercury arc in the various mercury lines is determined by the pressure at which the lamp is operated. A mercury pressure of about 0.008 mm is needed to keep most of the energy in the 254-nm line, and the mercury vapor pressure is dependent upon temperature. The effects of temperature are evident in three important lamp characteristics—light output, starting the lamp, and color.

Light output. The mercury pressure of highly loaded, 1500-ma lamps is partly controlled by lamp design. In the T-12 type (tubular,

12/8-in. diameter) the cathodes extend into the tube so that a mercury-condensation end chamber exists behind a shield. A lamp of non-circular cross section, such as the Power Groove, uses two points of constriction near the center of the lamp for mercury control. European lamps use a protrusion near the center of the lamp for this purpose. In every case the result is that part of the bulb wall operates at a lower temperature than the rest of the tube, thereby providing a mercury condensation point. Maximum light output will depend on keeping the mercury condensation point at the optimum temperature. Tube-wall temperature is controlled by the ambient temperature (fig. 2.15) and by the air flow over the lamp (fig. 2.16). Luminaires for lighting buildings usually rely on convection currents to cool the fixtures, but in plant-growth-chamber applications where the lamps are closely packed, often only 3 mm apart, considerable air movement is necessary.

A transparent barrier between the lamps and the growing area is used in most plant-growth rooms. A separate lamp loft is not essential, however, and many excellent plant-growth chambers have been built without one. Some loss of light is inevitable with the barrier. The light depreciation caused by the barrier increases with age, especially when the barrier is made of plastic. If outside air is used for lamp cooling, the barrier must be cleaned regularly. The use of a barrier is perhaps best justified by the reduction in vertical temperature gradient in the plant-growing area. Morris (1957) has calculated that a single-layer barrier reduces conducted and convected heat from lamps to growing area by as much as 47% and reduces radiated heat by 49%. The barrier also contributes to stable light output when plant-growing areas are operated at extreme temperatures.

Whenever a barrier is used, some kind of lamp-temperature control system is necessary. The most common method is ventilation of the lamp loft with air from the space outside the chamber or in some instances with air ducted in from outdoors. In either case the location of the growth chamber can have a marked effect on light output because of the temperature sensitivity of the lamps. If the chamber is placed in an air-conditioned space at 24–25 C, the air moved through the lamp loft will probably result in nearly optimum conditions for the 1500-ma, T-12 lamp but will be somewhat cool for the Power Groove, provided the air flow is satisfactory. When the chamber is located in an uncontrolled

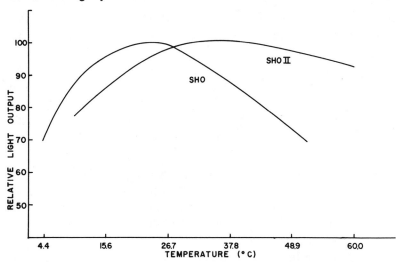

FIG. 2.15. Effect of ambient temperature on the light output from regular (SHO) and from high-temperature (SHO-II) 1500-ma, T-12 fluorescent lamps (Westinghouse Electric Co., *Bull.* A-8365A).

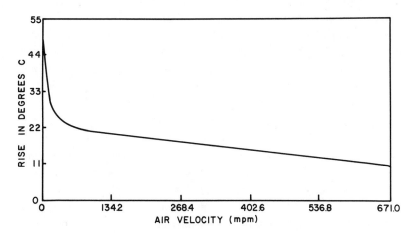

FIG. 2.16. Effect of air velocity on temperature of PG-17, 1500-ma fluorescent lamps (General Electric Co.).

space like a headhouse, however, lamp behavior can be a problem. During the summer the air will usually be too warm, but its temperature will fluctuate from day to day and even during the course of a single day. Light intensity will follow the temperature fluctuations, and constant light conditions can no longer be assured. In the winter the ventilation air may be too cold, and light output will again decrease. Considerable temperature control could be obtained in cool or cold weather by decreasing the air velocity or by mixing some discharge air with the cold inlet air to maintain a desirable temperature, but this is rarely done.

The most efficient system for lamp-temperature control is to equip the lamp loft with a separate heat exchanger of sufficient capacity to maintain the air passing the lamps at nearly optimum temperatures. The main problem with the closed system is that it must be designed correctly. Some systems use evaporatively cooled water as a coolant, and rate of flow can then become a critical matter, especially if the piping system uses a crossover connection on the suction side of the pump between supply and return lines. At the NCSU phytotron, inadequate water supply to such a system caused a 35–40% decrease in light output because of high lamp-loft temperatures. The problem was solved only when the phytotron staff redesigned the system.

Since the mercury condensation points are at the ends of the T-12 lamp but at the center of the Power Groove, air flows for the two types of lamps must be in opposite directions. This seems to be a point that some growth-chamber manufacturers fail to appreciate, since they imply that any kind of 1500-ma lamp can be used equally well in their chambers. A lamp loft designed for PG-17 lamps simply does not work effectively for the T-12 types (fig. 2.17).

Optimum air velocity for the growth-chamber lamp loft (whether for ventilation air or for recirculated air in closed systems) has not, to my knowledge, ever been determined experimentally. Since high temperatures are usually more of a problem than low ones, extrapolations from available data (fig. 2.16, *IES Lighting Handbook* 1966, General Electric 1967) would suggest that 50–80 m/min is required. Yet it is not at all unusual to find an air flow across the lamps of no more than 15 m/min. Thus, many growth chambers are constructed without adequate air velocity for the lighting system, without considering lamp design, and without any means of adjusting ventilation air temperature or velocity as a func-

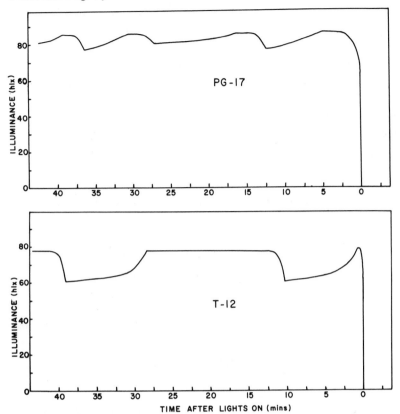

FIG. 2.17. Effect of fans cycling on/off as a function of lamp-loft temperature on the light output of PG-17 or T-12, 1500-ma fluorescent lamps.

tion of seasonal changes in the supply-air conditions. As a result, light output is not consistent and may be well below the level that could be produced.

Since light output is a function of mercury vapor pressure, which varies with tube-wall temperature, lamp design can cause a greater or lesser susceptibility to temperature. For example, smaller-diameter lamps like the T-10 have the same losses as the PG-17 but distributed over a smaller surface area. Thus the lamps tend to run hotter and respond faster to a change in temperature. As the tube mass and surface area become larger the lamp is less susceptible, and responses to temperature changes are slower. As a result, the optimum ambient temperature for the PG-17 is higher than that of the T-10, and less cooling is necessary.

When lamps operate at higher than optimum temperature, the number of premature failures increases dramatically. One type of 1500-ma lamp develops air leaks in a substantial percentage of lamps when operated at 35 C, a symptom that almost never occurs with lamps operated at 25 C. At the higher ambient temperatures the spring tension in the lamp holders seems to deteriorate more rapidly, with a corresponding increase in lamp failures due to arcing and heating at the poor contacts.

Certain fluorescent lamps have been designed for high-temperature operation and may prove useful for controlled-environment rooms where the lamp area temperature cannot be maintained satisfactorily. Some types use mercury amalgams to extend the practical temperature of the lamp. Others, such as the Sylvania Gold Band, use a strip of a rare metal that has a high affinity for mercury. The Westinghouse SHO-II uses an indium collar around the stem that supports the cathode. This latter lamp is reported to have an optimum operating temperature of about 40 C, as compared to the 25-C optimum of the regular series 1500-ma lamp. The effect of temperature fluctuations on the light output of these lamps should be less than on the regular series because of the much flatter temperature-response curve (fig. 2.15).

The loading of the lamp loft and the proportion of the energy that penetrates into the growing area are functions of lamp type as well as total number of watts. For example, the Lucalox lamp is the most efficient producer of light but it also causes a greater heat load in the growing area than does a fluorescent source (fig. 2.18). With a ventilated system, careful consideration would usually be given to the portion of lamp heat exhausted into an air-conditioned building, but conducted and convected heat from lamp lofts using a heat exchanger is often ignored. Yet the amount of heat lost through the highly conductive aluminum that makes up the cooling module is so great that even a few rooms can overload a zone air-conditioning system, despite the fact that the lamp loft includes a cooling coil. The aluminum sides of the lamp loft remain 10 C hotter than the ambient temperature when such a system is working

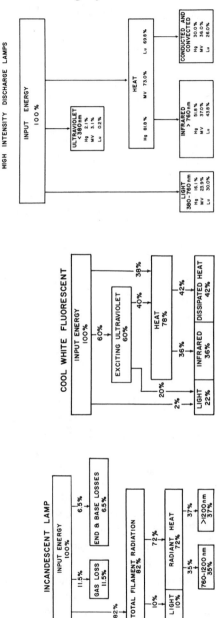

FIG. 2.18. Energy distribution from an incandescent-filament lamp, a fluorescent lamp (*Bull.* TP-111-R), and three kinds of high-intensity-discharge lamps (General Electric). *Hg* = mercury vapor; *MV* = multivapor; *Lu* = Lucalox.

correctly and can reach 20–30 C higher than ambient if the flow of cooling water through the coils is restricted.

Light output of HID lamps is not affected greatly by the ambient temperature. The main function of a cooling or ventilating system for HID lamps would be to maintain the lamp-loft temperature close to that of the growing space to minimize the vertical temperature gradient.

Starting the lamp. As the ambient temperature is reduced, fluorescent lamps become more difficult to start. Hard starting may become a problem, at about 10 C, depending on the quality of the ballast. For example, special ballasts can be obtained to ensure starting at temperatures as low as −28 C. In addition to hard starting, low temperatures can also cause appreciable blackening of the tube, often near the middle of the lamp, due to mercury condensation where one part is cooler than the rest. Very high temperatures are also reported to cause difficult starting (*IES Lighting Handbook* 1966).

Under conditions normally encountered in plant-growth-chamber applications, temperature would have little effect on the starting of HID lamps. Low temperatures can cause the HID lamp to start more slowly and, if it is consistently operated under conditions which result in long warm-up times, the lamp life as well as light output can be adversely affected.

If an HID lamp is turned off, it will not restart immediately as will fluorescent and incandescent lamps. The HID lamp must cool enough to reduce the vapor pressure to a level where the available voltage can restrike the arc. Cooling time depends mainly on installation factors such as ambient temperature and air velocity. Under any given set of conditions the metal halide lamp will generally require more time for restarting than either the mercury or sodium lamp.

Temperature and lamp color. The spectral energy distribution of the fluorescent lamp is the net result of the two or more phosphors plus the mercury arc. The various components do not react in the same way as the temperature of the lamp rises, and the larger contribution from mercury arc usually results in a shift toward green. The amount of the color shift is about the same as the maximum difference between lamps because of manufacturing variations (*IES Lighting Handbook* 1966), but if all the lamps in a growth chamber make such a shift simultaneously, the result may have biological significance.

Ballasts

A fluorescent lamp is started by applying sufficient voltage across the cathodes to strike an arc. As the current in the arc increases, the resistance decreases; and an unprotected lamp would continue to draw current until it destroyed itself. The purpose of the ballast is to limit lamp current.

The simplest ballast is merely a coil of wire placed in the circuit so that the induction limits the current. In modern lamps, however, the ballast is somewhat more complicated. A transformer in the ballast increases line voltage to provide the higher starting voltage. A transformer also decreases line voltage to provide the low voltage for the cathode heaters. A two-lamp ballast system contains a capacitor to correct the power factor (see Allphin 1959 for a discussion of the power factor) and another capacitor to suppress radio interference. All the components are placed in a metal box filled with potting compound, which reduces ballast noise and helps dissipate ballast heat.

Ballast life should be about 12–15 years when the lamp is operated 10 hr a day at the proper voltage and temperature (Elenbaas 1959, *IES Lighting Handbook* 1966). Unfortunately, premature ballast failure is a problem of frequent concern to the operator of a plant-growth chamber. The manufacture of ballasts is a highly competitive business, and some producers are not using high-quality components. Many ballasts, of course, give excellent service and provide reasonably well regulated current to the lamps. For adequate performance, only ballasts which feature CBM (Certified Ballasts Manufacturers) certification should be used in plant-growth-chamber applications.

Most ballasts are designed to operate with a case temperature below 90 C. This limit assumes conditions that will keep the winding temperature below 105 C and the capacitor below 70 C. A 10° increase over 105 C will reduce the life of winding insulation by half. As the insulation becomes heated and carbonization occurs, the dielectric strength is reduced until finally an electrical short occurs between the coil turns or between coil and case. If the ballast mounting system provides marginal ventilation and heat dissipation, overheating and consequent ballast failure are certain to occur.

The compounds mainly used for internal heat transfer from the components to the ballast-case surface have a high coefficient of expansion and tend to flow at elevated temperatures. Inadequate heat transfer from the case to the surroundings frequently causes the filler to be forced from the can. Loss of filler results in a less effective internal heat transfer, and one or more components will fail as a result. Ballast failure due to improper mounting and inadequate ventilation seems to be especially prevalent with built-in, owner-designed growth rooms. In some cases ballast life has been so brief and failure so predictable that drip pans have been installed to collect the pyranol forced from the cans, yet no efforts have been made to improve ventilation. Premature ballast failure on many growth chambers may be less obvious because the ballasts last several years before failing. Few growth-chamber manufacturers and few persons using the chambers have collected and evaluated long-term maintenance information, so no one really knows whether full ballast life is being obtained.

In lighting-fixture applications, ballasts are mounted with as much surface as possible against the metal of the fixture, which is in direct contact with the surrounding air. A similar technique is sometimes used by plant-growth-chamber designers, with varying degrees of success.

In some cases the ballasts have been mounted against the lamp loft, where heat from the lamps is conducted to the ballasts. The designer was apparently under the mistaken impression that by simply obtaining a good metal contact he was duplicating the methods used in commercial luminaires. The safest system is to mount the ballasts on rails away from the wall, with long axis vertical; a hinged cover should be placed over the entire ballast board, and forced ventilation should be no less than 60 m and preferably 90 m per minute. As with lamp ventilation in plant-growth chambers, however, no experimental evidence is available on optimum air flow for ballast cooling.

Some ballast manufacturers strongly recommend placing a fuse in line with each ballast. Although the fuse does not necessarily prevent ballast failure, the disagreeable odor and dripping potting compound can be avoided. Most manufacturers produce ballasts with built-in thermal protectors, and some use thermosetting resins instead of pyranol as a potting compound. Unfortunately, few growth-chamber designers use any of these protective devices, and they are rarely specified by the user of the chamber.

In a growth-chamber system the ballast should be placed outside the plant-growing area and outside the lamp loft. In commercial chambers the ballasts are usually mounted on the outside of the side or end wall, where they may or may not be enclosed in a cabinet. With small chambers no electrical problems are encountered in placing the ballasts outside; as room size increases, lamp-to-ballast distance may be great enough to require special precautions to avoid excessive voltage drop in the cathode circuit, a condition that will cause starting failures and short lamp life. Conductor resistance should be limited to 0.2 ohm for 1500-ma lamps. As distance between lamp and ballast increases, wire size must therefore also increase. For example, at 4.57 m a No. 18 (1.0 mm) wire can be used, but at 9.14 m the wire size must increase to No. 14 (1.6 mm). If a single wire is used for a pair of lamps, a No. 14 wire would have to be used at 4.57 m and be increased to No. 12 (2.0 mm) at 9.14 m.

Voltage drop resulting from the remote placement of ballasts of HID lamps should not exceed about 3% of the nominal operating volts. Metal-halide lamp ballasts can be removed about 24 m with No. 14 wire and up to 60 m with No. 10 wire. High-pressure sodium lamp ballasts can, if properly selected, be removed up to 10 m. New and improved ballasts are being designed for these light sources, however, and more remote locations may be possible.

The ballast system can, of course, be set up as separate components (Herrick and Wenner 1962). Since the low (3.2-v) voltage required by the cathode heaters is the major problem in remote placement of 1500-ma ballasts, the transformer for the cathode heater would need to be mounted in a wireway at the ends of the lamps. The remainder of the components could be several hundred feet away if necessary. An additional advantage of the separate components is that remote placement requires only one wire for every two lamps instead of the three per lamp for the commercial ballast. A commercially made lighting system was based on the circuitry of Herrick and Wenner (1962), utilizing heavy-duty, carefully sized components mounted separately in a well-ventilated cabinet. The result was a system that provided excellent current and power factor control with very low maintenance requirements. An imitator of the original system has failed to provide adequate cooling ventilation, and premature failure of components occurs just as it would with commercial ballasts.

Intensity Control

Programming the light as a smooth curve is difficult and therefore expensive. It is possible, using Herrick and Wenner's (1962) circuit, to incorporate a system of high-power, silicon-controlled rectifiers (SCRs) to create dimming by phase control. At very low light levels the a-c dimming system causes ripple or spiraling, a condition in which the brightness of the lamp varies from end to end in a cyclic fashion. This problem can most likely be overcome electronically, but such a system has never been developed.

A direct-current system for lamp operation is available commercially, using current as a controlled variable. Lamp flicker and spiraling are eliminated, lamp wiring is simplified, and ballasting can be remote. The current, and consequently the light level, can be related to a low-level command signal. A signal generator could create nearly any lighting program including square, sawtooth, or sine wave pulses from durations of fractions of seconds to several hours. The system is expensive and currently costs as much as the commercial growth chamber on which it might be installed.

Economical light-intensity programming thus seems to be reduced to using a variable number of lamps. Many growth-chamber manufacturers offer a 1/3, 2/3, 3/3 lamp-switching program, but in my opinion the steps are too broad. The lowest permissible light should be determined by the fewest lamps that will retain a good horizontal uniformity. Carpenter et al. (1965) have shown that with 1.52-m lamps in a chamber that apparently had a floor about 1.37×1.54 m, uniformity was not affected appreciably when 6 or more, equally spaced lamps were used in each program step. They also suggest that this number of lamps could be economically dimmed with a saturable reactor. The minimum number of lamps per step would have to be determined for each chamber design because room dimensions, degree of wall reflectance, and electrical factors would influence the result. Tests at the NCSU phytotron showed that with a room of 2.44 × 3.66 m, containing a total of 84 fluorescent lamps, uniformity of illuminance and balanced electrical phase and power factor required steps of 12 lamps each.

Adjusting intensity by removing lamps instead of switching them out of the circuit is satisfactory only when the lamps are removed as pairs matched to a ballast. If one lamp is removed from the two-lamp ballast circuit, the cathodes of the remaining lamp continue to heat, and premature ballast failure can result.

Lumen Maintenance

The amount of light produced by a fluorescent lamp decreases as the lamp is used, because the ability of the phosphor to convert the 254-nm radiation into visible light is slowly reduced (Lowry 1948, Allphin 1959, Elenbaas 1959, Carpenter and Moulsley 1960, *IES Lighting Handbook* 1966). The reduction in output with time is a function of the ratio of arc power to phosphor area. Thus lumen loss from highly loaded lamps like the 1500-ma types used in growth rooms is much greater than from the more lightly loaded ones commonly used in building lighting. Theoretically, with equal arc power, the larger diameter lamps like the PG-17 would have better lumen maintenance than the T-12 sizes.

Probably 10% of the initial illuminance is lost during the first 100 hr of use. This interval is usually referred to as the stabilization period, and no light measurements should be made until it is complete.

The most conservative figures predict a light loss of less than 20% after 4000 hr operation for a T-12, 1500-ma lamp. These data do not include the 10% lost during the stabilization period. According to manufacturers' data, therefore, a growth chamber with an initial, stabilized illuminance of 600 hlx would have about 480 hlx at the end of a year's operation at 12 hr per day. However, many growth-chamber users, and we are among them, have the impression that light output of lamps in controlled-environment rooms decreases at a considerably greater rate than manufacturers' data indicate. Manufacturers' tests of lumen maintenance are made under nearly ideal conditions, whereas lamps in growth chambers are much closer together than they would be in any other application, and they generally operate considerably warmer than the recommended temperatures. The contrast in the two sets of conditions is so great that one might conclude that the lamp environment has a definite effect on lumen maintenance although there are no data to support or refute such a conclusion. The rate of light loss is usually much greater than desirable for biological work, and a constant illuminance is usually maintained by periodic replacement of a portion of the lamps. Carpenter, Mousley, and Cottrell (1964) worked out a lamp-changing system that could be used as a pattern. The NCSU phytotron tries to balance labor and lumen maintenance by replacing the oldest one-third of the lamps whenever light intensity drops below 430 hlx.

The gradual reduction in light from HID lamps is principally the result of electrode emission material coating the walls of the arc tube. Thus lumen maintenance of HID lamps should be and is generally supposed to be superior to that obtained with fluorescent lamps. Manufacturers' data indicate, however, that light output of metal-halide lamps decreases about 20% after 4000 hr, about the same as with fluorescent lamps. Sodium-lamp output is reported to decrease only 10% during half the rated 20,000-hr life, considerably better than most other light sources.

Lumen maintenance of HID lamps depends in part on the number of starts per hour used but also on electrical factors such as wave form, lamp watts, and current. Maintained light output may even be affected by lamp position; for example, metal-halide lamps give the best lumen maintenance when used in a horizontal position. Thus, in practice, lumen maintenance may fail to match manufacturers' data.

Lamp Life

The life of lamps is defined as the number of hours they may be used before 50% of them burn out. Rated life is of no real importance with fluorescent lamps used in plant-growth chambers, because the effective life of the lamp ends whenever the lumen output drops below the minimum level required by the biologist.

Rated life of the incandescent lamp, however, is a good measure of the effective life in plant-growth applications. Efforts to increase the rather short life of the incandescent lamp have led to use of 130-vac lamps on 120-vac supply. Lamp life increases considerably but with a significant loss in light output (fig. 2.19) and a shift in the red–far red ratio (fig. 2.20). Recent tests with extended-service lamps, which have a rated life of 2500 hr and about the same lumen output as the 130-vac lamps, show that 25% of the latter were lost while no burnouts occurred

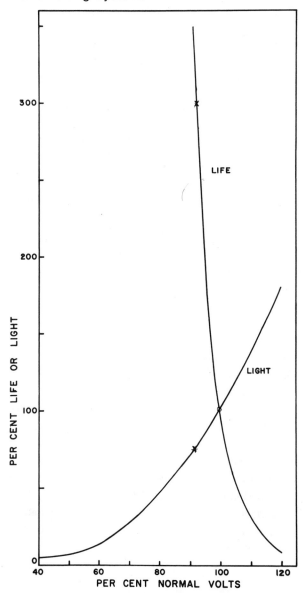

FIG. 2.19. Effect of operating voltage on life and on light output of incandescent-filament lamps.

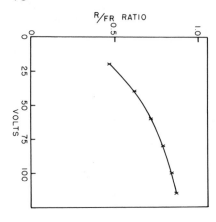

FIG. 2.20. The red—far red ratio emitted by a 120-v, 300-w incandescent-filament lamp operated at various socket voltages (Downs et al. 1964).

with the extended-service lamps. Additional tests using 105-w krypton-filled incandescent street lamps, which have a rated life of 12,000 hr and about the same lumens as the extended-service lamps, indicate they may be very useful where the lamp loft can dissipate the additional heat produced.

Photoperiod Control

In many growth-chamber applications the photoperiods are kept constant for many days, and the use of ordinary 24-hr time switches is adequate. Photoperiod lengthening is often desired without any significant increase in total radiant energy, so the incandescent lamps should be controlled by a separate time clock from the fluorescent ones.

Certain experimental objectives require more elaborate photoperiod or light-source programs. A system described by Downs and Bailey (1967) allows the insertion of various repeat-cycle time switches that will operate for any period set by a pair of 24-hr cycle timers. Carpenter et al. (1965) report on a controller with five modes of operation that could be used to program either the incandescent or some of the fluorescent lamps or some of both types.

Wall Reflectance

A discussion of growth-chamber lighting would be incomplete without including wall reflectance. Plant-growth chambers can be found that use mill-finish aluminum, stainless steel, aluminized foil, specular

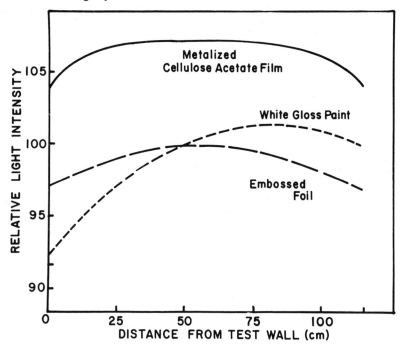

FIG. 2.21. Effect of wall surface on light uniformity (Carpenter and Moulsley 1960).

aluminum, and, of course, white paint. The prospective buyer of a plant-growth room need decide, however, only on the relative merits of white paint, aluminized foil, or specular aluminum (fig. 2.21). Very few data are available (Carpenter and Moulsley 1960, Carpenter et al. 1965), but it appears that specular aluminum would provide a somewhat higher reflectance than white paint. Moreover, and perhaps more importantly, the vertical gradient and the decrease in intensity as the wall is approached seem to be less with specular aluminum.

Aluminized Mylar and other foils are fragile and easily torn. Perhaps the torn area could be cut out and replaced with a patch of new material, but the high maintenance factors seem to make this material undesirable. A specular aluminum, such as Alzak, is quite resistant to corrosion, but salt sprays and splashed nutrient solution will spot it unless cleaned off frequently. Certain chemicals, including some of the currently used fogging pesticides, will cloud the specular aluminum, with a dramatic loss of reflectance. Specular aluminum is relatively hard but can still be scratched, and the scratches cannot be repaired. White paint is usually damaged by the same materials that would damage specular aluminum. Fogging and spotting would not be as readily detected, but the light loss would probably be about as great.

Generally, loss of reflectance, whether due to corrosion of specular aluminum, yellowing of white paint, or accumulation of dirt on any surface, is a slow process and may not be noticed by the investigators using the equipment. Periodically, therefore, the reflectance should be checked. Each time that illuminance or irradiance measurements are made, it would take little extra time to record a reading with the meter at right angles to a test section of wall, at a distance of about 30 cm. Room and wall illuminance will have a constant ratio as long as the reflectance is not changed.

Super-high Light Intensities

The fluorescent-incandescent system can only produce a maintainable light intensity of about 485 hlx. Some biologists believe that double this intensity is necessary for many kinds of research, especially with plants like corn, milo, and cotton.

When any lighting system other than the accepted fluorescent-incandescent one is desired, costs often rise many times higher than the real cost differential in the hardware and lamps. The step rise in cost is probably attributable to "engineering time"—a catchall phrase that more and more people believe increases the probability that the system will fail to work satisfactorily. Little engineering has really been done, however, with any HID lamp installations in high-intensity plant-growth chambers, and it seems unlikely that a really good system will be attained until such research is conducted. High-intensity-discharge lamps act as point sources and are very amenable to manipulation by reflection optics. In theory, therefore, the HID system can provide uniform illuminance clear to the walls in chambers of any shape or size. Since the HID lamps can be manipulated to provide a broad source effect, the vertical gradient should be no greater than with fluorescent systems.

Composition of the Atmosphere

The CO_2 content of a plant-growth chamber is rarely measured and even more rarely monitored on a continuous basis. Maintenance of CO_2 levels is more than important; it can often be essential for development of normal plants (Raper and Downs 1973, Raper et al. 1973). Corn can pull the CO_2 level in the growth room down from the 400-ppm ambient condition to approximately 80 ppm within 30–40 min after the light period begins. Tobacco uses enough CO_2 to bring the level from 400 ppm down to 180–200 ppm.

Makeup Air

Most growth rooms have some kind of makeup air system that is supposed to prevent this CO_2 depletion, yet the amount of air supplied is rarely reported. In almost all cases, makeup air systems fail to maintain ambient CO_2 conditions. Moreover, the presence of these devices has misled investigators into believing that a real function is being performed. Apparently no biological study has been made to determine the amount of makeup air required to prevent CO_2 depletion, although Morse (1963) has estimated it. He calculates that if the CO_2 consumption is 5 g hr^{-1} m^{-2} and if a 10% reduction is allowed, then 1,520 l. of air per minute per square meter of growing space would be required. In a 1.22×2.44-m growth room, about 4,524 l./min would be needed, and in a 2.44×3.66-m room, about 13,574 l./min. The ceiling height of many growth rooms is about 2.1 m, so an air exchange rate of about 0.7 times per minute is necessary by Morse's calculations. Yet, in one set of specifications that I have seen, the makeup air system on rooms of 2.44×3.66 m was specified at 705 l./min, a number supposedly based on the supply systems of commercially manufactured growth rooms. Actually, the installed system changed only 2% of the room volume per minute instead of the 3.7% that would have been changed by the specified 705 l./min. Larger blowers were installed to bring makeup air exchange to 12% of the room volume per minute. The fivefold increase in makeup air made only a 25% improvement in the level of CO_2 depletion caused by tobacco plants.

Considering the large amounts of air required, the use of makeup air to maintain ambient CO_2 conditions does not seem practical.

CO_2 Injection

At the concentrations of interest in plant-growth-chamber applications, bottled CO_2 gas is the most practical means of maintaining a constant CO_2 level. Control can be manual, through periodic adjustment of flow meters to balance the CO_2 supply against values measured by any convenient method. Automated CO_2 systems can use an infrared gas analyzer to operate meter relays that open or close solenoid valves in the CO_2 supply line. Bailey et al. (1970) suggest that proportional control could be obtained with three-way proportioning valves and appropriate electronic operators.

Since the plant load in terms of photosynthetic leaf area changes slowly, under constant environmental conditions the least complicated method of CO_2 control is satisfactory. Manual adjustment of a flow meter two or three times a week can easily maintain a 350–400 ppm concentration of CO_2, the adjustment being determined by the reading from an infrared gas analyzer that may or may not be used to record data continuously. The flow diagram of the CO_2 control system used at the NCSU phytotron (fig. 2.22) illustrates such a practical system that has an extremely low maintenance factor.

In greenhouses, CO_2 control can be more complicated because of changing light intensities. Pettibone et al. (1970) report on a light-modulated CO_2 control system. The various time constants set up by valve opening and closing rate, duration of sampling period, and especially greenhouse response time to an increase in CO_2 create a time lag that causes some fluctuation about the set point as the rate of CO_2 utilization changes with light intensity. Although Pettibone and co-workers present no data to indicate how constant a CO_2 level can be maintained with their system, it seems likely that a light-modulated controller would be an ideal method to use in greenhouses.

In completely air-conditioned greenhouses like those found in most phytotrons, normal CO_2 levels can usually be maintained by makeup air from outdoors. The large volume-to-leaf-area ratio of greenhouses allows CO_2 levels to be maintained with a smaller percentage of new air than would be required in plant-growth chambers.

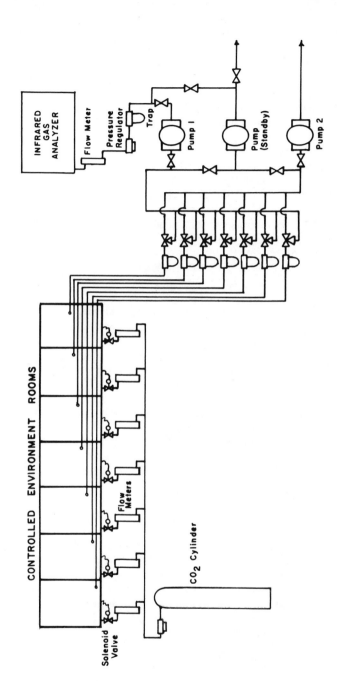

FIG. 2.22. System by which CO_2 is measured and controlled in a number of plant-growth chambers in the Southeastern Plant Environment Laboratories.

References

Allphin, W. 1959. *Primer of Lamps and Lighting.* New York: Chilton Co.

Anderson, D. B. 1936. Relative humidity or vapor pressure deficit. *Ecology* 17:277–82.

Bailey, W. A., H. H. Klueter, D. T. Krizek, and N. W. Stuart. 1970. CO_2 systems for growing plants. *Trans. ASAE* 13:263–68.

Blackman, G. E., and J. N. Black. 1959. A further assessment of the influence of shading on the growth of different species in the vegetative phase. *Ann. Bot.* 23:51.

Bohning, R. H., and C. A. Burnside. 1956. The effect of light intensity on rate of apparent photosynthesis in leaves of sun and shade plants. *Amer. J. Bot.* 43:557–61.

Carpenter, G. A., and L. J. Moulsley. 1960. The artificial illumination of environmental control chambers for plant growth. *J. Agr. Eng. Res.* 5:283–305.

Carpenter, G. A., L. J. Moulsley, and P. A. Cottrell. 1964. Maintenance of constant light intensity in plant growth chambers by group lamp replacement. *J. Agr. Eng. Res.* 9:60–70.

Carpenter, G. A., L. J. Moulsley, P. A. Cottrell, and R. Summerfield. 1965. Further aspects of the artificial illumination of plant growth chambers. *J. Agr. Eng. Res.* 10:212–29.

Downs, R. J., and W. A. Bailey. 1967. Control of illumination for plant growth. In *Methods in Developmental Biology,* ed. F. W. Wilt and N. K. Wessels, pp. 635–45. New York: Thomas Y. Crowell.

Downs, R. J., W. T. Smith, and G. M. Jividen. 1973. Effect of light quality during the high intensity light period on growth of plants. Paper no. 73-4525, winter meeting, American Society of Agricultural Engineers, Chicago, Ill.

Downs, R. J., K. H. Norris, W. A. Bailey, and H. H. Klueter. 1964. Measurement of irradiance for plant growth and development. *Proc. Amer. Soc. Hort. Sci.* 85:663–71.

Ellenbaas, W. 1959. *Fluorescent Lamps and Lighting.* New York: Macmillan.

Friend, D. J. C., J. E. Fischer, and V. A. Helson. 1963. The effect of light intensity and temperature on floral initiation and inflorescence development of Marquis wheat. *Can. J. Bot.* 41:1663–74.

Fuchs, M., and C. B. Tanner. 1965. Radiation shields for air temperature thermometers. *J. Appl. Meteorol.* 4:544–47.

Gaastra, P. 1959. Photosynthesis of crop plants as influenced by light, carbon dioxide, temperature and stomatal diffusion resistance. *Mededel. Landbouwhogesch. Wageningen* 59:1–68.

Gaastra, P. 1964. Some comparisons between radiation in growth rooms and radiation under natural conditions. *Phytotronique* 1:45–53. Paris: Editions du Centre National de la Recherche Scientifique.

General Electric. 1967. *Fluorescent Lamps*. Bull. TP-111-R, Large Lamp Division. Cleveland, Ohio: General Electric Co.

Herrick, P. R., and R. E. Wenner. 1962. Omnibus of special circuits for fluorescent lamps and lighting control. Paper presented at the winter meeting, Electrical Engineers, New York, N.Y.

IES Lighting Handbook: The Standard Lighting Guide. 1966. 4th ed., ed. J. E. Kaufman. New York: Illuminating Engineering Society.

Jividen, G. M., R. J. Downs, and W. T. Smith. 1970. Plant growth under high intensity discharge lamps. Paper no. 70–824, winter meeting, American Society of Agricultural Engineers, Chicago, Ill.

Lowry, E. F. 1948. A study of fluorescent lamp maintenance. *Illum. Eng.* (N.Y.) 43:141.

McCree, K. J. 1972. Test of current definitions of photosynthetically active radiation against leaf photosynthesis data. *Agr. Meteorol.* 10:433–53.

Meijer, G. 1971. Some aspects of plant irradiation. *Acta Hort.* 22:103–6.

Morris, L. G. 1957. Some aspects of the control of plant environment. *J. Agr. Eng. Res.* 2:30–43.

Morse, L. G. 1957. The design of growth rooms. In *Control of the Plant Environment*, ed. J. P. Hudson, pp. 139–56. London: Butterworth Scientific Publications.

Morse, L. G. 1963. Some recent advances in the control of plant environment in glasshouses. In *Engineering Aspects of Environment Control for Plant Growth*, pp. 62–95. Melbourne, Australia: CSIRO.

Müller, D. von. 1951. Analyse der Stoffproduktion von Gerste. *Bodenkultur* 5:129–35.

Parker, M. W. 1946. Environmental factors and their control in plant experiments. *Soil Sci.* 62:109–19.

Parker, M. W., and H. A. Borthwick. 1949. Growth and composition of Biloxi soybean grown in controlled environment with radiation from different carbon arc sources. *Plant Physiol.* 24:345–58.

Pettibone, C. A., W. R. Matson, C. L. Pfeiffer, and W. B. Ackley. 1970. The control and effects of supplemental carbon dioxide in air-supported plastic greenhouses. *Trans. ASAE* 13:259–63.

Plapper, H. 1972. "Phytosystems" for scientific plant research. In *Phytotronique* 2:110–20, ed. P. Chouard and N. de Bilderling. Paris: Gauthier-Villars.

Raper, C. D. 1971. Factors affecting the development of flue-cured tobacco grown in artificial environments. III. Morphological behavior of leaves in simulated temperature, light durations and nutritional progressions during growth. *Agron. J.* 63:848–52.

Raper, C. D., and R. J. Downs. 1973. Factors affecting the development of flue-cured tobacco in artificial environments. IV. Effects of carbon dioxide depletion and light intensity. *Agron. J.* 65:247–52.

Raper, C. D., W. W. Weeks, R. J. Downs, and W. H. Johnson. 1973. Chemical properties of tobacco leaves as affected by carbon dioxide depletion and light intensity. *Agron. J.* 65:988–92.

Senn, H. A., D. P. Anderson, and L. C. Anderson. 1965. *Manual for Investigators*. Madison: Univ. of Wisconsin Biotron.

Thomas, M. D., and G. R. Hill. 1937. The continuous measurement of photosynthesis, respiration and transpiration of alfalfa and wheat growing under field conditions. *Plant Physiol.* 12:285–307.

Thomas, M. D., and G. R. Hill. 1949. Photosynthesis under field conditions. In *Photosynthesis in Plants*, ed. J. Franck and W. E. Loomis, pp. 19–52. Ames: Iowa State Univ. Press.

Went, F. W. 1957. *Environmental Control of Plant Growth*. Chronica Botanica, vol. 17. New York: Ronald Press.

Withrow, A. P., and R. B. Withrow. 1947. Plant growth with artificial sources of radiant energy. *Plant Physiol.* 22:494–513.

3 ENVIRONMENTAL MEASUREMENTS

The literature clearly shows that instrumentation and methods of measurement vary so drastically in plant-growth-chamber and greenhouse applications that comparisons of controlled-environment facilities and the data obtained from them are difficult if not impossible. One growth-chamber manufacturer, for example, claims a light level of 538 hlx, whereas a second manufacturer using the same number, type, and spacing of lamps can apparently provide only 377 hlx. It is therefore hardly surprising to find that, reporting the same environmental conditions, one investigator may produce a plant 25–30% larger per unit of time than a second investigator. The reported differences in plant growth are not due to green-thumbism but simply to the fact that one or more of the environmental conditions were different.

Our ability to control the environment cannot exceed our ability to measure it. The reason we wish to control and measure the various environmental factors is to determine their effect on biological behavior, yet many of the measurements used to evaluate environmental effects are simply not valid. Since the degree of environmental control and the interpretation of experimental results rest so heavily on the sensing and readout system, measurement needs to be discussed in detail.

Air Flow

Air velocities in plant-growth applications are usually low, often about 25 m/min and rarely exceeding 46 m/min. The rate usually specified is 30 m/min and is based on plant-behavior data (fig. 3.1). The most satisfactory and relatively economical measuring device for these low velocities is the hot-wire anemometer. The basic arrangement consists of a fine wire with a large temperature coefficient of resistivity, heated by an electric current. Air passing the wire causes a decrease in wire temperature and consequently a change in wire resistance. The resistance changes

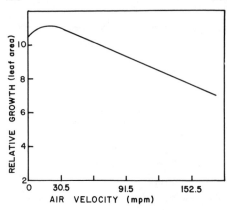

FIG. 3.1. Effect of air velocity on relative growth rate of leaf area (Morse 1963).

are then used as a measure of flow velocity. A single-wire system is directional, and response can change as position changes. With multiple-wire arrangements, a nondirectional system can be obtained (fig. 3.2).

Plants probably sense air movement in a nondirectional manner, so a multiple-wire anemometer would provide the most realistic data. The directional anemometer is very useful, however, for evaluating air-duct velocities, the uniformity of the growing area, and subsequent balancing of the space.

Direction

Plant-growth chambers have been constructed with flow from bottom to top, top to bottom (the so-called reverse air flow), and even from side to side. Air-flow measurements in an empty room would probably show bottom-to-top and horizontal flow methods to be close to laminar-flow systems. When plants are placed in the growing area, the laminar flow is broken up and considerable variation can be introduced. For example, one can calculate that when plants in 10-cm pots are placed in the growing area, pot-to-pot and wall-to-wall, the free area is reduced to the space between the pots, or from 3 m² to about 0.28 m². The increased static pressure could reduce air flow by its effect on fan characteristics, and certainly the velocity will increase from 30 m/min to as much as 300 m/min near the top of the pots. Since air velocity in the bottom-to-top air-flow system can be influenced by plant loading, it should be measured with plants in place at the beginning of the experiment when the plants are small, and at the end when the plants are large.

The reverse air system is turbulent, and an average flow of 30 m/min will produce very brief, random fluctuations that range from 15 to 45 m/min. The average air flow around the most photosynthetically active plant leaves remains near the optimum design level and is much less influenced by the plant load.

Measurements of air flows at plant level in controlled-environment facilities almost always should be recorded with the investigator removed from the space, the doors closed, and the room in a normal operating condition.

Temperature

Temperature transducers for measurement operate on the same principles as those already described for control. Cost, response speed, reproducibility, mass or volume of the sensor, and location are some of

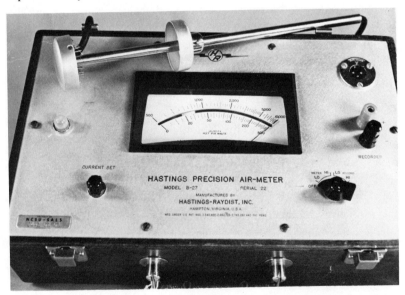

FIG. 3.2. Omnidirectional multijunction, hot-wire anemometer for measurement of air flow (Hastings-Raydist, Inc., Hampton, Va.).

the important factors to consider in plant-growth-chamber applications. Ideally, of course, response of the transducer to a change in conditions should be instantaneous. Since it is not, the response speed of the sensor will determine the degree of transient change that can be detected.

There seems to be little agreement on the kind of measurements that would be most representative of plant temperature. The degree to which a plant leaf will follow a transient temperature change cannot be determined without a method of measuring the plant's temperature. The response speed of some plants is probably about the same as that of a hygrothermograph. Others, especially those with broad thin leaves, may respond much faster and follow transient changes more closely.

How do we measure temperature as the plant senses it? Attempting to build a thin, flat sensor to simulate a leaf would seem doomed to failure because of the difficulties of matching absorption, reflection, and transmission characteristics, plus the fact that it would have no cooling effect from transpiration. The most widely accepted method seems to be the use of very small thermocouples pressed against or inserted into the leaf. The thermocouples must be selected for low thermal conductivity, which means that copper should not be used. Chromel-alumel is recommended by Trickett (1963) as being the best compromise for leaf temperature measurements. Investigators using this method claim that very small thermocouples do not absorb the large amounts of infrared transmitted by the leaf and therefore provide a realistic measurement.

Burrage (1972) uses five No. 40 thermocouples in series to get a measure of the average leaf temperature. Average temperature measurements can be made from the output of several thermocouples in parallel or in series connection. Parallel operation requires the same length of extension wire in each thermocouple circuit or some method of equalizing the resistances of the individual circuits. In series-connected thermocouples, the output is the sum of the individual outputs divided by the number of units,

$$E_T = \frac{E_1 + E_2 + \ldots + E_n}{n}$$

Extension wire polarity must be observed, and all intermediate junctions must be at the same temperature as the refrence junction.

Infrared thermometers provide noncontact measurement of surface temperature, and some of these instruments function in the ambient temperature range. The sensing system may include a multijunction thermopile, thin-film thermistor pairs, a bolometer, and probably silicon cells. These devices work very well when the emissivity, the rate of heat loss per unit area, of the body being measured is many times greater than the amount of infrared reflected from the surface. Plant emissivity unfortunately is very low in comparison to the amount of infrared reflected. In order to use infrared thermometers for measuring plant temperature, one must devise a method to separate the reflected and emitted components so that only the emitted one is measured.

The fact remains that universal agreement on a standard method of measuring plant temperature has not been reached. Moreover, in agriculture, in ecology, and in greenhouse horticulture, temperatures are reported as air temperature. Much of the biological data concerning temperature effects has thus been obtained using meteorological techniques, either with the sensors in a standardized weather shed or in some other housing that allows air movement over the sensors while shading them from radiant energy. It is my contention that air temperature, measured as accurately as possible, in a manner available to most investigators, should be a basic environmental parameter reported in all studies using controlled-environment facilities.

Temperature information from a sensor exposed to high light intensities is subject to the physical conformation of the sensor; therefore the measuring system should be shielded from radiant energy. Just as with the control transducers, temperature-measuring sensors must also be shielded from surrounding objects that absorb and reradiate to them. Moreover, the enclosure must be aspirated because errors up to 13 C can occur in unaspirated ones (Morris 1963). If the sensors are kept in aspirated enclosures, air temperature can be measured equally well by different investigators using different types and sizes of thermocouples, transistors, or resistance elements. Other types of sensors, properly calibrated, can give equally accurate measurements but may not be as sensitive to transient changes. We prefer to use No. 24 (0.51 mm), type T (copper constantan) thermocouples with a 0.5-mm, welded bead junction. Smaller thermocouples and small thermistors would be equally satisfactory.

A properly designed, aspirated housing undoubtedly allows the most accurate measurement of air temperature. Air flow through the housing should be standardized, probably at about 50 m/min. The best material for the housing seems to be specular aluminum (Fuchs and Tanner 1965). If an aspirated sensor housing is impractical, the return air duct could be used, provided there is no appreciable temperature rise during the passage of the air from plant level to sensor (fig. 3.3).

Air temperature alone is not a completely satisfactory measurement but, used along with a description of the light source, it will at least enable different investigators to discuss biological temperature effects in the same terms. More complete environmental information could be obtained by measuring the radiant temperature as well. The radiant temperature is a measure of the heating effect of the light source. A valid estimate of the heating effect might be obtained with two thermocouples or thermistors, one placed inside a table-tennis ball painted flat black and one in another ball coated to give a high reflectance. When they are exposed to radiant energy at plant level, the difference in the temperatures of the two balls is an estimate of the radiant temperature. Read et al. (1963) discuss this type of measurement but use thermocouples embedded in chrome-plated and blackened steel balls. Radiometers can also be used to separate the long-wavelength radiation from visible light. This is done by painting one radiometer receiver white and the other black. White, magnesium oxide paint reduces the sensitivity to visible light by nearly 85% without affecting the long-wavelength sensitivity. The long-wavelength effect is determined as the difference between net radiation and visible radiation (Funk 1959). Regardless of the method, the interest is in the ability of the 800–3000-nm radiation from the light source to raise the temperature of plant and substrate.

Air temperature is a valid measurement, of course, only when the biological material is in the air. Aseptic culture of tissue, fungi, or algae in flasks, mycorrhizae research where seedlings are grown in bottles or fruit jars, or boll weevil studies in which the animal is inside the boll represent temperature environments that may be considerably higher than the surrounding air. In such cases the temperature near the organism must be measured, and in a typical fluorescent-incandescent–lighted plant-growth room it will be about 5–6 C higher than the air. The temperature inside the container will vary much less than the surrounding air and will change more slowly in response to a day/night temperature alternation.

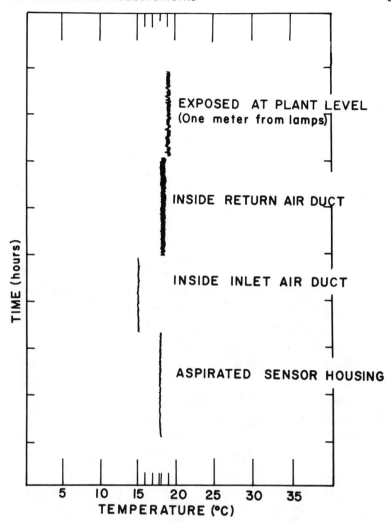

FIG. 3.3. Effect of location on the response of a thermocouple temperature sensor.

I do not contend that air and radiant temperature measurements are "valid" in the sense that they represent leaf, root, or substrate temperatures. I argue only that they form part of a basic set of reproducible environmental measurements that should be reported to describe the experimental conditions.

Relative Humidity

Hair and hairlike elements in hygrothermographs (fig. 3.4) are probably the most common means of relative-humidity measurement. Gold–lithium chloride cells respond faster, however, and are used in many growth-chamber applications (fig. 3.5). These types of RH sensors can be calibrated against an aspirated psychrometer, a device that uses matched thermometers, one of which is enclosed in a wet wick. The dry-bulb and wet-bulb temperatures are then used to obtain relative humidity. Ventilation rates of 150–300 m/min are usually needed to obtain stable results. Since RH sensors respond faster in a moving air stream, the

FIG. 3.4. The hygrothermograph (Bendix Corp., Baltimore, Md.).

problem decision is the same as with temperature measurements: whether to use an aspirated housing or to use normal growth-chamber air movements. Matthews (1965) shows that a lithium chloride hygrometer required 16.4 sec to indicate 90% of the change from 75 to 29% RH at 31 C when ventilated at a rate of 98 m/min. At 183 m/min, 90% response was obtained in only 7.2 sec. Because of the close relationship between temperature and relative humidity, it would seem reasonable to measure humidity at the same location as temperature, although the lower air velocities recommended for temperature transducers would not produce the maximum response rate from lithium chloride sensors.

Dew-point temperature—the temperature at which water vapor condenses for a given state of humidity and pressure as the temperature of the vapor is reduced (*ASHRAE Handbook of Fundamentals* 1967)—is considered by some investigators to be a more reliable and realistic system than relative humidity. For a given amount of water in the air, say .014 kg per kilogram of dry air, the relative humidity would change from about 88% at 21 C to about 46% at 32 C. The dew point relative to that amount of moisture would stay constant at 19 C. Relative humidity therefore fails to provide realistic information concerning the vapor pressure or the actual amount of water in the air. Dew-point temperature, on the other hand, because it depends only on the amount of water in the air, results in a more absolute measurement with direct inference on the evapotranspiration potential of the air.

A number of instruments are available for dew-point measurements. Some types use lithium chloride sensors that measure the vapor saturation of a saturated solution of the salt. Others use an optical system that determines the temperature at which water condenses. This latter type (fig. 3.6) uses a cooled metal surface in the gas stream. When the temperature of the metal is cooled to the dew point, condensation occurs. When an equilibrium exists between the water condensing and the water evaporating, the metal surface is at the dew-point temperature. This temperature is indicated by a precision thermistor which forms part of a bridge circuit. The presence of the condensate is detected by a light beam and photocell. The photocell changes resistance as a function of the light incident upon it, and the output of a photoconductive bridge is directed to a thermoelement in the mirror. The optical-thermal feedback control thus adjusts the mirror temperature to provide a constant reduction in reflec-

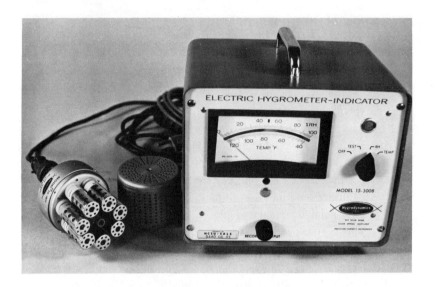

FIG. 3.5. Gold–lithium chloride relative-humidity transducer and readout system (Hygrodynamics, Silver Spring, Md.).

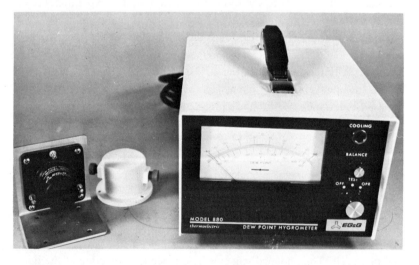

FIG. 3.6. Optically sensed, thermoelectrically cooled, condensation dew-point hygrometer (E. G. & G., Boston, Mass.).

tance, which requires a constant amount of condensate, only attained at the dew-point temperature.

Radiant Energy

The light used for plant growth has been measured in almost every conceivable manner (Bickford and Dunn 1972). The term "light" implies that we are interested in that portion of the electromagnetic spectrum with wavelengths between 380 and 780 nm; by definition, light is that portion of the spectrum that stimulates human vision and extends over the spectral region seen by the human eye (fig. 3.7). Therefore a method of measuring light was developed that was based on the sensitivity of the eye to both brightness and spectral quality. The illumination units are based on the luminous flux per unit of solid angle from a standard source referred to as a candle. A standard candle is said to emit 4π lumens (lm) or 1 lm per steradian, and the luminous intensity at a distance of 1 ft would be 1 footcandle (ft-c). At a distance of 1 m, the luminous intensity over 1 m^2 would be a meter-candle or a lux. Thus 1 ft-c would equal 10.76 lux. In the International System of Units (SI units) all illuminance measurements are based on lux (Allen 1968). At high light levels the values are large and are measured in hectolux (hlx), where 1 hlx = 100 lux.

One type of illumination meter is made from barrier-layer cells that generate a voltage when exposed to radiant energy. The barrier-layer or photovoltaic cell (fig. 3.8) is generally made with crystalline selenium as the semiconducting medium. Light striking the barrier layer causes a voltage to be generated between the positive base plate and the negative transparent layer. The short circuit current is proportional to the area of the cell and increases linearly with illumination. Output current deviates from linearity, however, at very high light levels or if only a part of the cell surface is illuminated. The spectral response of such cells is somewhat different from that of the human eye, and therefore color-correcting, photopic filters are added. A light meter used in plant-growth rooms should be hermetically sealed to protect the cell against the high humidity and temperature that may be found in the controlled space; measurements should be made rapidly to prevent internal heating of the cell. If the high

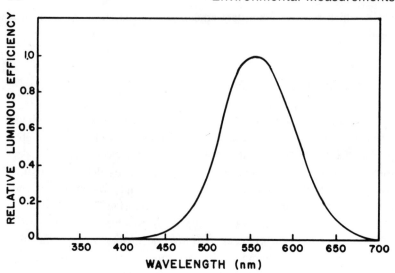

FIG. 3.7. Spectral response of the human eye.

light levels of the plant-growth chamber should heat the cell above 49 C for an appreciable time, it could be permanently damaged. Optimum selenium-cell temperature is about 25 C. Note that this is the internal temperature of the cell, not the ambient temperature in which it is used.

The barrier-layer cell and its accompanying microammeter are inherently inaccurate. The meter alone can easily have an error of ±6–8% at any point above 1/4-full scale. It is not at all unusual for two apparently identical meters to disagree by 30% when used to measure illuminance levels above about 215 hlx. Illumination meters can be used satisfactorily for plant-growth lighting only if they are frequently recalibrated.

Light measurements become involved with considerations of the inverse-square law and the cosine law. The inverse-square law states that the intensity, I, of the source varies inversely with the square of the distance, d, of the surface from the source. The inverse-square law is not valid when the distance is less than five times the largest dimension of the source. In many plant-growth chambers the entire ceiling is the light source, and the distance-to-source ratio is 1:1 or less. Therefore, vertical movement of the measurement position records only a small change in illuminance. For example, in a 0.91×1.22-m chamber with a 0.91×1.22-m light source, measurements at $d = 4$ and $d = 3$ differ by only about 12% instead of the 43% predicted by the inverse-square law. High-intensity-discharge lamps are much more representative of a point source, but when reflectors are used the effect is that of a broad source, so a doubling of distance reduces illuminance to 60% instead of the calculated 25% of the initial value.

The receiver of the barrier-layer cell is usually flat and sometimes is set below the edge of the mounting ring. Light with a high angle of incidence is reflected by the glass cover and the cell surface, and the rim of the mount may obscure part of the receiver. The cosine law states that the illuminance of a surface varies as the cosine of the angle of incidence ($E_2 = E_1 \cos \theta$). The loss due to failure to include light with a

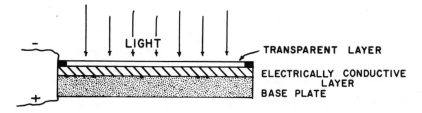

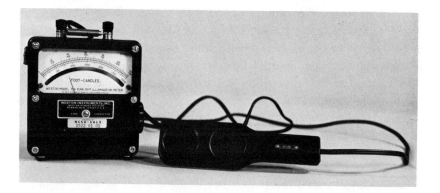

FIG. 3.8. The photovoltaic or barrier-layer cell. A schematic drawing and a typical readout system (Weston Instruments, Inc., Newark, N.J.).

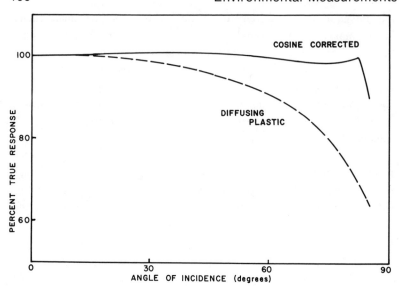

FIG. 3.9. Cosine responses of a cosine-corrected sensor and a sensor with a diffusing plastic cover (Biggs et al. 1971).

high angle of incidence can be as much as 25% (*IES Lighting Handbook* 1966). Therefore these instruments, and indeed all light-measuring systems used for plant growth, should be equipped with a cosine correction device. Unfortunately the trend is to place a piece of plastic over the receiver and call it cosine corrected without actual determination of the degree of correction obtained (fig. 3.9). In only a few cases have manufacturers attempted to determine the cosine correction and furnish such data with the instrument or upon request.

Illumination meters are also available that use phototubes or photomultipliers. These are classed as photoemissive transducers because the light incident upon a cathode liberates electrons which move to an anode and thus form an electric current. The cathodes are coated with an alkali metal that has a small work function but that also has increased yield in certain spectral regions. In other words, they are sensitive to the wavelength of the incident radiation. With the proper choice of cathode surface, these transducers can be calibrated in terms of lux, and excellent systems are available commercially (appendix 3).

Illumination units are based on the spectral sensitivity of the human eye, as well as its brightness sensitivity. The eye sees some wavelengths better than others, and illumination meters are carefully watched to provide the same spectral response. Herein lies the error of using the hectolux as a unit of light for plant growth; the plant's spectral sensitivity is not the least similar to that of the human eye. In addition the sensors are calibrated from incandescent lamps having a color temperature of 2854 K, which means that an error is produced whenever light is measured from lamps at a different color temperature. Nevertheless we are likely to continue to use the illumination unit, despite irrelevance and high errors of measurement, because instrumentation is relatively inexpensive and because a large number of investigators have illumination meters. Moreover many biologists are able to relate illuminance to plant growth more readily than any other type of units. Illumination units are of course perfectly valid as a relative measurement for comparison of the light produced by the same kind of light source, from one experiment to another, or between experiments in different laboratories.

Total energy is usually measured in the form of irradiance, the radiant flux per unit of area. Some investigators seem to prefer irradiance measurements in terms of watts per square meter, which can be easily converted to other units (100 w m^{-2} = 10 mw cm^{-2}) for comparison with results of photobiological research. Meteorological units like the calorie cm^{-2} min^{-1}, sometimes called the Langley/min, also appear in the biological literature, as do ergs cm^{-2} sec^{-1} and Joules sec^{-1}.

Total energy measurements include infrared to rather long wavelengths, often from 300 to 3000 nm or more. Since radiation longer than about 800 nm plays no known role in photobiology, one could argue that such measurements are no more valid for plant growth than is illuminance. The heating effects of the long-wavelength radiation should be evaluated, but as radiant heat rather than light or energy for photochemistry.

Thermopiles, bolometers, pyrheliometers, and pyranometers all measure total energy and have been described in detail many times (Forsythe 1937, Withrow and Withrow 1956, Lion 1959). All are thermal-radiation transducers consisting essentially of a blackened receiver, often a thin metal strip, which absorbs radiation so that the temperature of the element is increased. The temperature rise is measured and used to indicate the radiant energy.

Thermopiles like the Hewlett-Packard radiant flux meter are

composed of very small thermocouples. One set of junctions, usually blackened, is exposed to radiation and the other is protected and isolated from it. The bolometer receiver changes resistance as a function of the radiant energy input. The sensor forms one leg of a Wheatstone bridge, whose imbalance is proportional to the incident energy. The Smithsonian pyrheliometer measures the solar radiation–induced rise in temperature of a blackened receiver by means of a precision thermometer, whereas the Ångstrom type is electrically compensated.

Pyranometers of the Eppley type are differential thermopiles with the hot junction a blackened receiver and the cold junction highly reflective. This type of pyranometer measures total sun and sky radiation from 280 to 2800 nm over a range of 180 C, but it is not cosine corrected. Other types of pyranometers such as those manufactured by Lambda Instruments and Yellow Springs Instruments (see appendix 5) have good cosine response and are sensitive to radiation between 300 and 2000 nm. These latter units use silicon-cell sensors which, unlike thermopiles and bolometers, do not have a flat spectral response. They are calibrated against a thermopile type, usually the Eppley pyranometer, for sun and sky radiation. When silicon-cell or photodiode instruments are used to measure the radiant energy emitted by any other kind of source, an error occurs due to that spectral sensitivity. The error for solar radiation measured by a silicon cell under a plant canopy or under incandescent lamps would be small. Under fluorescent light sources, one unit had an error of 43% whereas another, recently calibrated radiometer produced an error of only 12%. These types of radiometers therefore should be recalibrated for use in plant-growth rooms.

Irradiance between certain wavelengths, usually arbitrarily selected as 400 and 700 nm, has been used as an indication of the photosynthetically active radiation, or PAR. Unfortunately, leaves do not have a constant response to all wavelengths over this range and consequently do not match the nearly flat spectral response of the measuring device (fig. 3.10). Therefore PAR measurements must contain an error factor, whose magnitude would depend upon the spectral emission of the light source but would not exceed 5% (Biggs et al. 1971). The PAR can be measured as irradiance in watts per square meter or as quantum flux in microEinsteins per square meter seconds. McCree (1972) suggests that the quantum flux measurements contain the least error and are probably

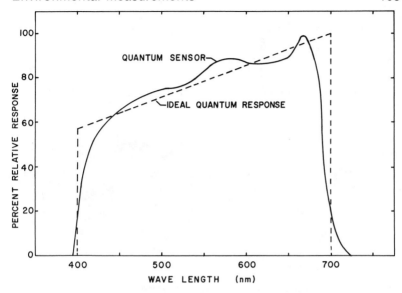

FIG. 3.10. Spectral response curve of the PAR quantum sensor and the ideal photon sensor (Biggs et al. 1971).

close enough for practical purposes. The best argument for describing radiant energy in terms of incident photons is simply that absorption of light by a biological system is a quantum process.

Commercially made instruments for measuring PAR are available at reasonable cost (fig. 3.11), and the selenium cell of illumination meters can be converted to approximate an equal quantum response for the 400–700 nm region by the addition of a Wratten 85C filter (Federer and Tanner 1966). It seems certain that PAR measurements will find increasing application as biologists begin to correlate photon density with plant growth. Figure 3.12 shows the relationship of illuminance and quanta for cool white fluorescent lamps and for high-pressure sodium lamps (Lucalox). This graph allows a comparison of quanta with a value that has greater familiarity, the hectolux.

PAR measurements can lead to erroneous conclusions regarding plant growth because no data are included to indicate the red–far red ratio that controls phytochrome. Thus, at equal PAR, abnormally slow repro-

FIG. 3.11. Multipurpose radiant-energy measuring system. Sensors are available to determine illuminance, irradiance, or PAR (Lambda Instrument Co., Lincoln, Nebr.).

ductive development could occur under lamps that lack sufficient radiation between 700 and 800 nm. It would be equally possible that some kinds of plants with high PAR and high far-red levels would elongate more than those grown with less PAR and less far-red. A better measurement would include wavelengths between 400 and 800 nm, or at least an estimate of the 700–800 nm radiation in addition to the PAR measurements. The Lambda Instrument Company has developed a sensor to supplement the quantum sensor in order to measure 700–800 nm radiation. This device is available as an addition to the equipment.

Table 3.1 illustrates the effect that the spectral quality of the light source has on the light measurement. Increasing the installed watts 38% by the addition of incandescent lamps raised the total energy 43% but the PAR only 14%. Far-red radiation increased 146%.

The most descriptive measurement of the energy available for plant growth is spectral irradiance over a wide range of wavelengths, at least 380–800 nm. Complete distribution curves of spectral energy provide adequate information for every purpose of comparing biological

TABLE 3.1 Effect of supplemental incandescent lamps on total irradiance

Light source	Total energy [a] (wm^{-2})	Far red, 700–800 nm [b] (wm^{-2})	PAR, 400–700 nm [b] (μE m^{-2} sec^{-1})
1600-w Metal halide	113	3.2	293
1600-w Metal halide plus 600-w incandescent	162	7.9	334

[a] Instrumentation from Eppley Laboratory Inc.
[b] Instrumentation from Lambda Instrument Co.

results obtained under various lighting situations. Total energy or total quantum flux can be obtained. Quantum flux at any wavelength can be calculated, and the strong morphogenic effect of the radiation between 700 and 800 nm is easily evaluated. Spectral irradiance completely describes the light source in terms of the kind and amount of radiation that reach the plants and takes into consideration barrier absorption characteristics, wall reflectance, and spectral shifts of lamps due to age, temperature, and voltage.

Spectral irradiance is measured with a spectroradiometer, usually in terms of microwatts per square centimeter per nanometer. Some types use a grating monochromator and S-1 phototube (Yocum, Allen, and Lemon 1964, Bulpitt, Coulter, and Hamner 1965), some use a wedge-interference filter (Norris 1964), and others use a number of fixed-wavelength–interference filters (Robertson and Holmes 1963). Commercial units of all types are available (see appendix 5). The grating-monochromator types are fine laboratory instruments but not really practical for many controlled-environment-room applications. The currently marketed version of the wedge-interference filter suffers from poor reproducibility of wavelength setting. Moreover, the wavelength scale is in 50-nm increments, which make it difficult to follow large changes in irradiance over a short portion of the spectrum and may cause some confusion of the resultant curve because of mercury or other emission lines. The fixed-wavelength spectroradiometer has the drawback of predetermined wavelength settings, which is offset to some degree by the distinct advantage of data directly, without the need for later alteration with correction factors. Furthermore, there is no limit on the number of filters. If each filter is electronically corrected, several paddles of a few filters each can be used, one paddle for each region of the spectrum.

The major disadvantage of a spectral-irradiance measurement is the cost of the instrumentation. Unlike McCree (1972), I believe that spectroradiometric measurements are practical in crop ecology, in agriculture, and especially in research that deals with controlled environments. I do agree with him that such measurements will not become common, but because of cost and maintenance rather than usefulness and desirability.

Perhaps the most surprising aspect of environmental measurement is the undying faith that so many otherwise astute investigators seem to place in their instrumentation. The reasons for this are not clear,

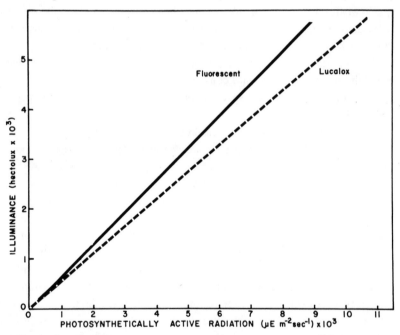

FIG. 3.12. Relationship between illuminance and photosynthetically active radiation for fluorescent and high-pressure sodium (Lucalox) light sources.

but I have repeatedly seen resesearchers assume that since a pen on a chart records 20 C, it must be so. Some investigators have reported illuminance values higher than the lamps are physically capable of producing, and others have noted relative-humidity control that can only be described as remarkable. Very few courses in environmental measurement are available, so perhaps the biologist's great confidence in his instruments can be explained by lack of opportunity to understand the intricacies of measurement. The problem is not confined to biologists, however, but also arises with people who should know better, such as manufacturers of environmental controls. When the NCSU phytotron was constructed, the representatives for the building-controls manufacturer proudly showed a greenhouse temperature recording that indicated ± 0 C variation. Investigation of this highly improbable situation showed that the thermocouple wires were attached backward on a recorder that had no off-scale movement for thermocouple failure.

Measurement is a science in itself and should be treated accordingly. Familiarity with the physical principles of the transducers is helpful, and an understanding of the fundamentals of the factor to be measured seems mandatory for valid data. If such knowledge is unavailable, old-fashioned common sense is very helpful in avoiding the more obvious errors.

Let us accept the fact that environmental measurements are relatively crude, even when correctly performed. Temperatures are rarely measured to 0.1 C, or energy to an accuracy of 0.1 μw cm^{-2}, and there is usually little reason for doing so. Yet the measured value should have some resemblance to the true one, and therefore instruments have to be kept in calibration. Merely because an instrument is new, it does not necessarily read the measured value correctly. In fact, a recently received digital thermometer had a persistent error of 1.5 C. All instruments need to be checked when received, then recalibrated at appropriate intervals. Those with mechanical linkages will need to be checked more frequently than the purely electronic types.

If we accept the need for frequent recalibration of measuring apparatus, the question that immediately arises is how to do it without elaborate equipment. Often the best method is to return the instrument to the manufacturer periodically; many manufacturers recommend such a procedure and suggest appropriate intervals. With most instruments the investigator can make checks to determine the accuracy of indicated values. An ordinary incandescent lamp operated at a known but reduced voltage and constant current can serve as a pseudostandard. A newly calibrated light-measuring instrument can be checked against such a source, then rechecked periodically to determine if the readings remain constant over time of use. Some companies that make light-measuring instrumentation can provide a calibration lamp for this purpose, and in many instances the system can be recalibrated by the investigator.

Temperature-measuring systems are no more reliable than light-measuring ones, although they are usually regarded with less suspicion. Yet the most commonly used systems, the hygrothermograph and the hydralic temperature recorder, need the most frequent attention. These instruments can at least be checked against a good laboratory thermometer, and electronic readout devices such as thermocouple potentiometric recorders can be checked by applying a known input voltage. Full instructions are usually given by the manufacturers of such instruments.

Reporting Environmental Data

The American Society of Horticultural Science has prepared a set of guidelines for reporting environmental data (Krizek 1970). Although all of the items would provide useful information, many are beyond the reach of most investigators. Even in phytotrons, where considerable effort is spent measuring the various environmental parameters, the guidelines could not be followed in their entirety. Tissue analysis, for instance, has been performed on a few species grown in the NCSU phytotron, but such data are not available for every type of plant material and variety studied. Moreover, a sample report including all the items listed required about four typewritten pages, probably more than most journals would consider reasonable.

The minimum requirements of the guidelines provide the necessary environmental data to enable others to reproduce or compare conditions and methods. Even these few items need not be discussed if they are not relevant to the biological experiment. H. A. Borthwick's Plant Physiology Laboratory reported perfectly reproducible results for years without owning any air-flow instrumentation.

I would propose the following approach to reporting environmental data:

The controlled environment facility. A brief description might say that the room is 2.44 × 3.66 m, with a 2.1-m growing height, specular aluminum walls, and a Plexiglas "G" barrier.

Light. The light source should be described, i.e., 1500-ma cool white fluorescent lamps and 130-vac frosted incandescent lamps in a 3:1 ratio by installed watts. The fluorescent lamp may be described even more precisely as FR96 T-12/CW/1500-135° or as F48 PG-17/CW.

Some measure of the amount of light at plant level should be given, such as 460 hlx or preferably 730 μE m^{-2} sec^{-1}, at 1 m from the lamps.

Instrumentation should always be described. If the receiver is not cosine corrected, the angle of view should be given.

Experimental conditions should be given in terms of the duration of the high-intensity light period and the photoperiod. A room may operate with 10 hr of high-intensity light and a 12-hr photoperiod obtained by extending the high period with 30 hlx (or 60 μE m^{-2} sec^{-1}) from 130-vac incandescent lamps.

Temperature. Air temperature should have been measured in a shielded, aspirated housing or in the return air duct. The report would include the measuring system and might be described as day/night temperatures of 26/22 C, with a variation about the set point of ± 0.25 C as measured with a type T thermocouple and potentiometric recorder. If the temperature is programmed diurnally, a copy of the recorded program should be presented.

When thermographs are used, the data need to be accompanied by a description of the way the instrument was shielded from radiant energy.

Relative humidity. In practice, RH is not reported unless it is of considerable importance to the experiment. Many investigators lack the instrumentation. Whenever possible, however, at least a general statement about relative humidity should be made. For example, dew-point or RH readings could be made just prior to watering and about 30 min afterward. Hygrothermograph data could be used as maximum and minimum RH, along with an approximate average.

Air flow. Direction of air flow and an estimate of velocity should be reported. The latter can be obtained from manufacturers' brochures or contractors' test data if instruments are unavailable.

Carbon dioxide. Since CO_2 is rarely controlled, the concentration should be measured when the plant load presents the maximum leaf area. A laboratory infra red gas analyzer can be used for this purpose. This is such an important consideration that if measurements cannot be made, I recommend that the investigator at least provide an estimate of the total leaf area in the controlled-environment space.

Cultural practices. The cultural practices are reported in the usual way. Type and size of containers, substrate composition, nutrient solution or fertilizer and frequency of application would normally be included.

Additional information. The most useful extra data would be radiant temperature and spectral irradiance, followed by substrate and leaf temperatures.

The uniformity of the various environmental factors over the area in which the plants are grown would also be of interest, especially in new types of controlled-environment chambers for which such reports are not available. Estimations of uniformity are rather easy to obtain by simply moving the light meter or temperature sensor around the growing area. Precise measurements of uniformity require care, time, and remote readout systems to isolate the space from the presence of the investigator. Biological tests of uniformity are made frequently, since plant-to-plant variance or position variance is included in the analysis of many experiments.

References

Allen, C. J. 1968. Light metering goes international. *Light* 37:22–3.

ASHRAE Handbook of Fundamentals. 1967. New York: American Society of Heating, Refrigeration and Air Conditioning Engineers.

Bickford, E., and S. Dunn. 1972. *Lighting for Plant Growth*. Kent, Ohio: Kent State Univ. Press.

Biggs, W. W., A. R. Edison, J. D. Eastin, K. W. Brown, J. W. Maranville, and M. D. Clegg. 1971. Photosynthesis light sensor and meter. *Ecology* 52:125–31.

Bulpitt, T. H., M. W. Coulter, and K. C. Hamner. 1965. A spectroradiometer for the spectral region of biological photosensitivity. *Appl. Opt.* 4:793–97.

Burrage. S. W. 1972. Analysis of the microclimate in the glasshouse. In *Crop Processes in Controlled Environments*, ed. A. R. Rees, K. E. Cockshull, D. W. Hand, and R. G. Hurd. London: Academic Press.

Federer, C. A., and C. B. Tanner. 1966. Sensors for measuring light available for photosynthesis. *Ecology* 47:654–57.

Forsythe, W. E. 1937. *Measurement of Radiant Energy*. New York: McGraw Hill.

Fuchs, M., and C. B. Tanner. 1965. Radiation shields for air temperature thermometers. *J. Appl. Meteorol.* 4:544–47.

Funk, J. P. 1959. Improved polyethylene-shielded net radiometer. *J. Sci. Instrum.* 36:267–70.

IES Lighting Handbook: The Standard Lighting Guide. 1968. 4th ed., ed. J. E. Kaufman. New York: Illuminating Engineering Society.

Krizek, D. T. 1970. Proposed guidelines for reporting studies conducted in controlled environment chambers. *Hort. Sci.* 5:390.

Lion, K. S. 1959. *Instrumentation in Scientific Research*. New York: McGraw Hill.

McCree, K. J. 1972. Test of current definitions of photosynthetically active radiation against leaf photosynthesis data. *Agr. Meteorol.* 10:433–53.

Matthews, D. A. 1965. Some research on the lithium chloride radiosonde hygrometer and a guide for making it. In *Humidity and Moisture: Measurement and Control in Science and Industry*, ed. A. Wexler, vol. 1, pp. 228–47. New York: Reinhold.

Morris, L. G. 1963. Some recent advances in the control of plant environment in glasshouses. In *Engineering Aspects of Environment Control for Plant Growth*, pp. 62–95. Melbourne, Australia: CSIRO.

Morse, R. N. 1963. Phytotron design criteria engineering considerations. In *Engineering Aspects of Environment Control for Plant Growth*, pp. 20–39. Melbourne, Australia: CSIRO.

Norris, K. H. 1964. Simple spectroradiometer for 0.4 to 1.2 micron region. *Trans. ASAE* 7:240.

Read, W. R., D. W. Cunliffe, H. L. Chapman, and J. J. Kowalczewski. 1963. Naturally lit plant growth cabinets. In *Engineering Aspects of Environment Control for Plant Growth,* pp. 102–25. Melbourne, Australia: CSIRO.

Robertson, G. W., and R. M. Holmes. 1963. A spectral light meter: Its construction, calibration, and use. *Ecology* 44:419–23.

Trickett, E. N. S. 1963. Environmental measurements in a glasshouse. In *Engineering Aspects of Environment Control for Plant Growth,* pp. 196–211. Melbourne, Australia: CSIRO.

Withrow, R. B., and A. P. Withrow. 1956. Generation, control and measurement of visible and near-visible radiant energy. In *Radiation Biology,* ed. A. Hollaender, vol. 3, *Visible and Near-Visible Light,* pp. 125–258. New York: McGraw-Hill.

Yocum, C. S., L. H. Allen, and E. R. Lemon. 1964. Photosynthesis under field conditions. 4. Solar radiation balance and photosynthetic efficiency. *Agron. J.* 56:249–53.

4 BIOLOGICAL ASPECTS OF CONTROLLED-ENVIRONMENT ROOMS

A number of decisions must be made at the time plants are placed in the plant-growth room. Many of these decisions are not necessary in greenhouse or field applications and often create problems for the investigator. Day/night temperatures, relative humidity, carbon dioxide, illuminance, total light intensity, photoperiod, and even to some extent light quality must be set at levels predetermined by the investigator. Unfortunately, the biologist frequently has inadequate information on proper environmental conditions for his plant material and must base his initial work on environmental levels obtained from greenhouse or field studies. Since most controlled-environment rooms do not duplicate the variations and fluctuations of the field and greenhouse, initial work can be disappointing. The investigator may have to perform a number of preliminary studies to establish optimum conditions before a satisfactory research program can be started. Cultural practices used successfully in the field or greenhouse may also have to be modified, and new watering schedules and nutrient requirements established.

Temperature

For any particular plant species under a given set of conditions, there is an optimum temperature regime for maximum growth and development. The optimum regime for seedling growth may differ from that for plant growth, and still other optima may exist for flowering and for fruit development. Maximum growth, however, is not necessarily the same as optimum growth because the very rapidly growing plants may not develop normally. For example, tomato plants grow most rapidly with 27-C day and 20-C night temperatures, but the plants are ''spindly'' and ''etiolated'' (Went 1957). Sturdier, more desirable plants are obtained using 24-C day and 17-C night temperatures. Went also reports that when night temperatures are high, at 27 C, only 2 or 3 flowers are produced per inflorescence, whereas at 10 C the inflorescences branch and produce up to 50 flowers. Unfortunately, at 10 C the fruit set is poor because of empty or abnormal pollen grains. The best overall production of growth, flowers, and fruit of tomato plants was obtained with 24/17 C day/night temperatures, not the 27/20 C that produced maximum growth (Went 1957).

Went (1957) subscribes to the idea of thermoperiodicity, that the temperature during the light period should be different from that of the dark period for optimum plant growth and development. Certainly many plants seem to grow and develop better with a night differential (Kramer 1957, Went 1957, Bhatti 1971), and the magnitude of the day/night differential seems to depend greatly upon the species. Hellmers (1963, 1966) reports that woody plant species such as Jeffrey pine and red fir showed increased growth with increasing differences in day and night temperatures as large as 13 C, whereas coastal redwood grew best with only a 2.8-C differential. The day/night differential also appears to change as the plants grow. Went (1957) states that with an 18-C day, optimum night temperature for growth of chili pepper plants gradually decreased from 30 to 26 to 21 C, and after 3 months was 16 C. At a 27-C day temperature, growth rates were greater, and the optimum 16-C night temperature was reached after 1.5 months. A sliding optimum was also found for fruit formation.

Optimal temperature regimes have been reported for only a few plant species, so few that optimum temperature tables are not included in the *Biological Handbooks* (1962, 1964, 1966). What data are available can be so conflicting that they are useful only as guides. Such confusion is understandable when one considers sliding optima and the dependence of temperature optima for various physiological processes on other environment factors. For example, optimum temperatures are reported to vary with the light intensity (Went 1957, Hellmers 1963) and with CO_2 concentration (Klueter et al. 1971), and considerable differences may exist between varieties (Went 1957). Except for a very few examples (Matsui and Eguchi 1972), we have not reached the state of push-button or computer-programmed rooms in which the best conditions are automatically adjusted as a function of growth. Moreover, since many of the temperature requirements were established in low-light growth chambers or under the variable light and temperature of greenhouses, optimum conditions will need to be reaffirmed for each variety or cultivar grown in modern controlled-environment facilities at 430–538 hlx.

When deciding on a temperature regime to use in a plant-growth room, the investigator should be aware that the amount of heat supplied to plants in a controlled-environment facility may be deceptive. For example, tobacco plants in fields in the Raleigh, North Carolina, area receive average day/night temperatures of 29.5/19 C in June, but in the plant-growth room a pattern of 26/22 C is much closer to optimum at this stage of development. Few plants grow properly with growth-chamber day temperatures above 30 C, although they may grow rapidly. Tobacco plants grown on a 34/30 C day/night temperature cycle produced smaller but thicker leaves that were inrolled and savoyed. Anthesis occurred but few seeds were produced.

A constant temperature held throughout the light period and a somewhat lower but equally constant one during the dark period create a very abnormal situation. Yet this is the way in which most plant-growth chambers are operated and the way in which much of our knowledge of temperature effects has been obtained. While one might argue that such temperature data have little value for crop production or ecological studies, they at least provide a starting point in attempting to understand and predict the whole-plant response to temperature conditions. Only when such background information has been obtained is the investigator in a position to design experiments intelligently, using programmed, continuously variable temperature regimes.

Far more important than the diurnal program is a simulated seasonal progression of temperatures. Vernalization is a partial program of this type, forced upon the investigator by the stubborn refusal of many plants to develop satisfactorily without it. Raper (1971) has illustrated the simulated temperature progression in detail and shows that fieldlike tobacco plants are difficult to obtain in controlled-environment facilities without such a program.

Relative Humidity

As discussed earlier, relative humidity is perhaps the least important environmental factor to try to control. Of course, at low humidities a water stress may occur as transpiration increases (Pallas et al. 1965), or watering may be needed so frequently that it creates a nutrition problem. Some studies (Hoffman 1971) show that low RH will substantially reduce growth and fruit production, and Weibe and Krug (1973) report that calcium deficiencies occur in cauliflower plants grown in high humidities.

Studies dealing with evapotranspiration, air pollution, insect physiology, and pathological problems require RH control. Many of these studies are specific, and each RH requirement must be handled as it arises. Much pathological work requires very high humidities. Sporulation often will not occur below 95% RH, and spore germination or infection occurs best on wet surfaces.

For studies in which relative humidity is not a critical aspect of the program, a vapor pressure deficit equal to an RH of 70% at 25 C is satisfactory.

Carbon Dioxide

One can reasonably conclude that plants grown in most growth rooms are under a CO_2 stress during most of the light period. The plants themselves create the stress condition by absorbing CO_2 faster than the makeup air and room leakage can replace it. The effect is not as great as would be obtained with artifically induced CO_2 stress conditions, and the plants may be green and appear healthy. As a result the reduced leaf area, reduced leaf dry-weight, and perhaps lower reducing-sugar content may be erroneously considered the normal growth obtained in the facility.

Internodes of tobacco plants under depleted CO_2 conditions are much shorter than those of field-grown plants (Raper and Downs 1973). Epinasty of leaves during maturation, which is rarely observed in field culture, occurs in the depleted CO_2 conditions, especially under a simulated seasonal progression of temperatures. Use of supplemental CO_2 to maintain ambient levels at 350–400 ppm in the growth chamber prevents the epinasty and generally makes the plants more like "normal" field-grown ones both chemically and physically.

Corn, which theoretically could reduce the CO_2 level to zero, can in practice reduce it below 100 ppm. The corn remains green and continues to grow, and we have no data on the changes induced by the CO_2 stress. However, the corn-depleted CO_2 level may well be below or very near the CO_2 compensation point for many plants, and other species sharing the environmental space may suffer. Geranium, marigold, and no doubt many other plants placed in the same room with corn begin to turn yellow, and their growth is severely retarded. When removed from the CO_2 stress, these plants return to a normal color and growth rate. However, if they were left for a long enough period under depleted CO_2 conditions, it seems unlikely that they would recover fully.

Enhanced Carbon Dioxide

Raising the CO_2 level above normal ambient concentrations has been referred to as CO_2 enhancement, CO_2 enrichment, and CO_2 fertilization. Several investigators have shown that net photosynthesis increases when the CO_2 level is raised (fig. 4.1). Hesketh (1963) showed that the net rate of photosynthesis rose from 47 to 92 with corn and from 24 to 69 mg dm^{-1} hr^{-1} with tobacco when the CO_2 concentration was increased from 300 to 1000 ppm. Increased photosynthesis should mean increased total plant growth, at least part of which should be represented in increased yields. However, the results of practical applications of enhanced CO_2 techniques seem to be inconsistent and generally confusing. Some reports of tomato reproduction show a significant increase in yield only about 50% of the time (Kretchman and Howlett 1970), and the increase never exceeded 25%. Fruit set and fruit size rarely improved as much as 20%. Other reports (Gardner 1966, Wittwer 1966) indicate substantial increases in yield if CO_2 is added to tomatoes planted in November, when light intensities may be relatively low. In the Netherlands about 90% of the lettuce area is under enhanced CO_2 for spring crops, but use of CO_2 in the autumn is restricted because of high night temperature and low light intensity (Van Soest 1966). The increase in lettuce yield is reportedly substantial. Wittwer (1966) has obtained 88% increases with added CO_2, and by using nitrogen fertilization he increased the yields as much as 147%.

The increased growth resulting from enhanced CO_2 is often readily apparent (Wittwer and Robb 1964, Bailey, Krizek, and Klueter 1969, Kretchman and Howlett 1970), and data show an increase in root development (table 4.1) and in flower production (table 4.2). However, the

Biological Aspects

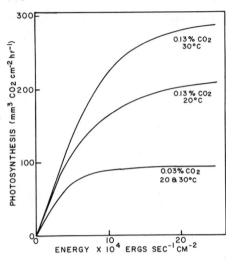

FIG. 4.1. Effect of light intensity and CO_2 concentration on photosynthesis of a cucumber leaf (Gaastra 1963).

response is often not as great as expected from photosynthesis data, and the effect frequently does not persist. Whittingham (1973) notes that although a very significant increase in dry weight of tomato plants occurred with enhanced CO_2, the effect slowly diminished with time. Tobacco behaves in much the same way, with the large effect of increased CO_2 concentration during seedling growth decreasing rapidly after transplantation. The interaction between CO_2 concentration and age or stage of development of the plant has not been thoroughly investigated, but it can be a major factor in the variability of results.

Tobacco, cucumber, pinto bean, gloxinia, and other plants under enhanced CO_2 conditions often seem to lose chlorophyll and become yellower than similar plants under normal CO_2 concentrations. Leaves may curl under at the edges, become savoyed and more leathery in texture. Hesketh (personal communication, 1972) noted the high starch con-

TABLE 4.1 Effect of enhanced CO_2 on root growth of Cherry Belle radish

CO_2 Concentration (ppm)	Fresh weight (g)	Dry weight (mg)	Diameter (cm)
300–400	3.1	203	1.7
1000–1200	11.8	681	2.8

TABLE 4.2 Effect of enhanced CO_2 on growth and flowering
of Happy Time White petunia

CO_2 Concentration (ppm)	Fresh weight (g)	Number of open flowers	Number of buds over 1 cm
300–400	20.7	0	<1
1000–1200	45.6	17	8

tent of cotton leaves when plants were grown at high CO_2 levels, and
Madsen (1968) reported starch eight times higher in tomato plants with
enhanced CO_2 while chlorophyll decreased from 1.5 mg/g to 0.7 mg/g
fresh weight. Measurements of tobacco leaves at 1200 ppm CO_2 revealed
72 mg of starch per gram of fresh weight while at normal levels only 7.6
mg/g fresh weight was obtained.

Several investigators have remarked that nutritional requirements
increase as CO_2 concentration is raised, but data are usually not provided.
Increased nitrogen levels reduce the starch accumulation and increase the
chlorophyll content in the leaves of plants grown at high CO_2, but with
tobacco appreciable increases in weight do not occur. Increasing the concentration of a balanced nutrient program can produce a marked increase
in fresh weight with enhanced CO_2; a doubled concentration was about
optimum for the plants tested (table 4.3). Radish root production, however, decreased with increased nutrient concentration, irrespective of the
CO_2 level (table 4.4).

Carbon dioxide enrichment is reportedly effective in greenhouses
during the winter when the light intensity is low (Hopen and Ries 1962,
Wittwer and Robb 1964). Whittingham (1973), who also found CO_2 en-

TABLE 4.3 Effects of nutrient concentration and CO_2 level
on fresh-weight production of Dare soybean (18 days after
planting)

Nutrient strength	Fresh weight (g)	
	Ambient CO_2, 300–400 ppm	Enhanced CO_2, 1000–1200 ppm
1/4	11.6	16.4
1/2	9.6	22.1
1	8.9	19.4

TABLE 4.4 Effects of nutrient concentration and CO_2 level on root growth of Cherry Belle radish (18 days after planting)

Nutrient strength	Root, fresh weight (g)	
	Ambient CO_2, 300–400 ppm	Enhanced CO_2, 1000–1200 ppm
$1/4$	11.6	22.5
$1/2$	9.3	19.9
1	6.4	16.7

hancement to be effective with low light levels, concluded that ''this can only be explained if enhancement results in a different metabolism as well as a change of rate.'' He further stated that he could not offer an explanation of how light, temperature, and CO_2 concentration mutually interact.

Carbon dioxide enhancement can definitely be used as a tool to increase growth and yields in controlled-environment facilities, but the results may not be as great as anticipated. Generally, plants grown in high-CO_2 atmospheres are more sensitive to changes in other environmental factors and in cultural practices. Insufficient research has been performed, however, to allow specific recommendations, so the investigator will need to plan on making some adjustments in procedures for each species being studied.

Light

Illuminance and Total Light Intensity

The level of illumination in the modern plant-growth room is relatively high but still amounts to less than half the maximum of sunlight on a bright summer day. However, if the total energy in the visible portion of the spectrum is integrated from all directions over the entire duration of the light period, the total energy of the growth room may be greater than that in the shaded greenhouse and nearly equal to that accumulated each day in the field. For example, solar illumination of a horizontal surface increases from zero to 1000–1300 hlx after a number of hours, then decreases again to zero. The illumination in a plant-growth room rises to 430–538 hlx in a few seconds and remains at that level throughout the light period. Moreover, a plant growing in the field or greenhouse receives only a small amount of light from the side, whereas sidelighting in the growth room can amount to 50% of the vertical illumination. Thus, as Gaastra (1964) illustrates, the total energy per day outdoors can be equaled or exceeded in the plant growth room on 70% of the days during the growing season. Gaastra is referring to photosynthetically active radiation (between 400 and 700 nm) and is not including the long infrared that makes such a large contribution to solar radiation but not to plant-growth chambers.

Although there are many research papers on the ways in which light intensity affects plant growth, many details remain to be examined and explained. For example, plants frequently produce more growth with a lower light intensity given for many hours than they would with a high intensity but shorter duration, the total energy per day being approximately equal (table 4.5). Some investigators (personal communications) have had difficulty in producing satisfactory tomato plants with long periods of high-intensity light, i.e., 15–16 hr, although the plants do well with 9, 10, or 12 hr. When unsatisfactory plants occur with long periods of high-intensity light, nutrition and CO_2 depletion may be involved. Each researcher will have to determine the optimum duration of high-intensity light for his own plant material and cultural practices. Moreover, the decision should be reached through experimental tests and not by the average day-length of the growing season or some duration that is supposedly "natural."

The biological merit of very high light intensities for controlled-

TABLE 4.5 Response of Cherry Belle radish and Fireball tomato plants to two relationships of light intensity and time

Light period (hr)	Illuminance (hlx)	Relative total energy	Fresh weight(g) radish root	Fresh weight(g) tomato top	Dry weight(g) tomato top
9+3 [a]	480	4320	9.0	29.8	1.97
16	247	3952	—	47.0	2.70
19	226	4294	15.8	—	—

[a] A 3-hr interruption of the dark period, from 11:00 P.M. to 2:00 A.M.

environment facilities is unproven. At illuminances of about 1000 hlx, radiant heat can become a serious problem and affect biological behavior, sometimes adversely (Biamonte 1972). Research on the benefits obtained from these light levels (Jividen, Downs, and Smith 1970) has as yet failed to show a satisfactory benefit-to-cost ratio; additional evidence established with proper removal of radiant heat needs to be obtained.

Light Quality

One would expect light quality to affect the efficiency of the high-intensity light period because of the spectral sensitivity of the photosynthesis system. One might also expect lamps matched to the action spectra of photosynthesis to be more effective than "white" lamps. In practice, however, the concept fails to fulfill expectations. For example, Went (1957) showed that a combination of red and blue lamps produced more dry weight than warm white ones. Note, however, that the data are dry weight per foot-candle. Since the foot-candle has a spectral sensitivity peaking in green, it would seem that a given foot-candle level of illuminance from red and blue would require more lamps than the same level from warm white lamps. The total energy between 400 and 800 nm must therefore be considerably greater, and increased growth might be expected.

When one uses equal installed watts, the red-blue combination is not as efficient as warm white light. Since in plant-growth chambers one can change fluorescent lamp types but can rarely increase the number of lamps, the only practical evaluation of various fluorescent lamp types is by equal installed watts. Theoretical evaluation of efficiency would have to be made on the basis of equal irradiance between 400 and 700 nm.

Plants certainly grow differently under light from the various regions of the spectrum (Arthur and Guthrie 1928, Funke 1931, Wassink and Scheer 1950, Meijer 1959), although "white" light usually produces plants that are more normal in appearance. The effect of various regions of the spectrum may be through the photosynthetic mechanism, but the photoreactions that regulate plant morphogenesis are also involved. Two of the most striking morphogenic photoreactions are the phytochrome system and the high energy reaction (HER). Some biologists believe the HER is a manifestation of the phytochrome system, others believe it to be a separate photoreaction; investigators shift from one view to the other as new data appear. Photomorphogenesis and the role of the HER and phytochrome have recently been discussed by Mohr (1972).

Phytochrome controls many aspects of the physiology of plants, including those grown in plant-growth rooms. Because of the ease with which the phytochrome system can be deliberately or inadvertently manipulated in the plant-growth room, an understanding and appreciation of the function and physiology of this ubiquitous pigment are essential. Phytochrome is a pigment that exists in two forms, a red-absorbing form, P_R, and far-red–absorbing form, P_{FR}. It is generally agreed that P_{FR} is the biologically active form. The balance or ratio of the two pigment forms largely determines many aspects of plant morphogenesis. Light with the proper spectral distribution, given at the right time, can control seed germination, internode elongation, flowering and fruit production, coloration of many plant parts, and dormancy.

Many plant-growth rooms contain both fluorescent and incandescent light sources. One of the major differences between them is the amount of far-red light produced, or perhaps more significantly, the ratio of red to far-red. A cool white fluorescent lamp has about a 5:1 R/FR ratio whereas the incandescent R/FR ratio is about 0.75:1. Increasing the total growth-chamber illuminance 10% by addition of incandescent lamps thus changes the red–far-red ratio from 5:1 to 2.3:1. This amount of change in the red–far-red ratio would cause the total biomass accumulation of Pollo corn to increase from 14.85 to 16.22 g (J. C. Stevenson, unpublished data), which is about the same benefit as reported by Went (1957) and by Parker and Borthwick (1949). In Stevenson's data the effect was obtained by adding the incandescent for only 8 hr of a 9-hr light period. The last hour before the dark period was fluorescent-lit only. In this way Stevenson avoided the very marked photomorphogenic effect of a decreased R/FR ratio at the beginning of the dark period (Downs, Hendricks, and Borthwick 1957).

The equilibrium of the phytochrome system at the end of a light period is one of the factors that determine plant behavior. When plants receive fluorescent light prior to darkness, the 5:1 R/FR ratio leaves phytochrome predominantly in the far-red–absorbing form. Internodes may be abnormally shortened, and in some plants flowering may be severely retarded (Downs, Hendricks, and Borthwick 1957, Downs, Borthwick, and Piringer 1958, Downs, Piringer, and Wiebe 1959). Sugar beets will probably not flower at all, and wheat, barley, millet, dill, and many other plants will flower very slowly.

Ending the day with incandescent light, on the other hand, leaves phytochrome in about a 1:1 ratio of red- and far-red–absorbing forms. Flowering may be accelerated, especially of long-day plants, but undesirable elongation can sometimes occur in other kinds of plants. Bush beans may assume a pole habit, and pole beans may begin to elongate precociously.

Pollen viability or production can also be influenced by the R/FR ratio at the beginning of the dark period. For example, elongation of the filaments of *Hyoscyamus niger* flowers seems to be inhibited by the presence of high levels of P_{FR} at the beginning of the dark period. The *H. niger* plants receiving fluorescent light prior to entering the dark period develop flowers in which the filaments fail to elongate; therefore no pollen is produced and few fruits are formed. The response is somewhat modified by temperature. At 21–24 C a complete loss of fruit set was observed, whereas at temperatures above 27 C some fruits formed regardless of light source (table 4.6).

All too often, light regimes are set up without consideration of the effect on the phytochrome system. In one example an investigator set up a program to begin the day with incandescent light, then give both

TABLE 4.6 Reproduction of *Hyoscyamus niger* as affected by the source of supplemental light

Supplemental light source [a]	Temperature (C)	Stem length (cm)	Days to anthesis	Fruit set [b] (%)
None	27	0.2	vegetative	—
	21	0.1	vegetative	—
Incandescent	27	42	27	66
	21	50	35	41
Fluorescent	27	34	36	12
	21	36	53	0

[a] Eight hours of supplemental light following 8 hr in the greenhouse.
[b] Twenty-five days after anthesis.

TABLE 4.7 Effect of light quality at the close of the light period on growth of tobacco seedlings

Variety	Light source [a]	Stem length (cm)	Fresh weight of leaves (g)	Stem diameter (cm)
Coker 319	fluorescent	13.0	25.0	4.1
	incandescent	21.0	30.3	9.7
Coker 254	fluorescent	13.0	30.6	5.2
	incandescent	21.3	33.4	10.3
Coker 298	fluorescent	10.6	27.0	3.6
	incandescent	19.2	29.1	9.6
NC-2326	fluorescent	10.9	28.2	4.6
	incandescent	15.2	30.0	7.4
NC-98	fluorescent	9.3	27.0	3.6
	incandescent	16.8	24.6	8.9

[a] For 0.5 hr at the end of 8.5 hr of fluorescent plus incandescent at 430 hlx.

fluorescent and incandescent, followed by fluorescent only during the final portion of the light period. Oat plants flowered more slowly than expected and failed to produce viable pollen. The problem was solved and normal flowers were obtained by ending the light period with incandescent light rather than fluorescent. Unfortunately, no further investigation of the phenomenon was made.

In another case, two growth chambers were set up with identical environmental conditions, but tobacco plants in one room produced abnormally short internodes. All environmental conditions were carefully checked out and proved identical in the two rooms until observations were made at the end of the day. Incandescent lamps went off 40 secs before the fluorescent ones in the room where plants had the short internodes, whereas the two light sources went off exactly together in the room with the more normal plants. Adjusting the light schedule induced greater internode elongation and the plants nearly recovered the loss in stem length. Results of an experiment to test the effect of the light source are shown in table 4.7.

Photoperiod

The importance of proper photoperiod selection cannot be over-emphasized. The physiology of vegetative plants is obviously different from that of plants in a flowering condition. The difference in metabolism begins during the flower initiation period, before any "visible," macroscopic evidence is apparent. Misleading and difficult to reproduce experimental results in studies using growth regulators or herbicide chemicals, mineral nutrition, or host resistance may result from differences in the physiology of vegetative or reproductive plants.

It is equally obvious that dormant woody plants are not physiologically the same as plants in a rapidly growing condition. The physiology of woody plants begins to change during the 30 or more days it takes the plants to become dormant. Since the change takes place slowly, it may not become apparent during the course of the experiment; or, by the time the decrease in growth is observed, much experimental time may be lost. Such a circumstance might happen with a nutritional study of a woody plant like the dogwood. The experiment may begin in the spring in a greenhouse, be conducted all summer, then in the fall show evidence of nutritional malfunctions as the plants stop growing and go dormant.

In the plant-growth chamber, increasing the day length not only establishes a photoperiod but also, if the total light system is used, increases the assimilatory period. The results of the two effects can easily be confused. For example, long days at high intensity light accelerate the flowering of many plants, such as flax, yet these plants do not show any true photoperiod response. In other words, plants like flax do not respond to interrupted dark periods or to extensions of the light period with low intensity light—conditions that induce a long-day response in photoperiodically sensitive plants.

True photoperiod control operates through the activation or deactivation of the active form of phytochrome, P_{FR}. Red radiant energy, free of far-red, is supposed to be the most effective region of the spectrum for controlling day length (Borthwick and Hendricks 1960), yet when "white" light is used, the incandescent lamp with its high far-red emission is usually a much more effective light source than the fluorescent lamp with its low far-red emittance (Vince and Stoughton 1957,

Downs, Borthwick, and Piringer 1958, Downs and Piringer 1958, Piringer and Cathey 1960).

Cultural Practices

We frequently hear the complaint that plants do not grow as well in the plant-growth chamber as they do in the greenhouse. Plants grown in controlled-environment rooms are reported to develop abnormalities such as twisted, malformed cotyledons, leaf epinasty, and nutrient deficiencies in soil known to be well fertilized. Nearly everyone who has worked with plant-growth chambers has experienced some kind of difficulty with the plant material, and many of these problems are related to cultural practices.

Went (1957) suggested that iron deficiency in corn resulted from low day temperatures, although "low" was not defined. Tobacco apparently becomes chlorotic at day temperatures of 7 C, and the chlorosis is accentuated by high night temperatures. Potato is said to show a leaf-edge burn similar to that produced by sodium or high salt toxicity when day temperatures exceed 27 C. Went (1957) reports that when night temperatures exceeded 25.5 C, tomato, chili pepper and potato became nitrogen deficient. Further studies indicated that the nitrate requirement varied directly with the night temperature and that at 12 C the nitrate could without harmful effect be reduced to one-third the amount required at 25.5 C.

Our own observations indicate that many cultural problems bear some relationship to the light intensity and, less strongly, to total energy. As the light intensity increases, nutritional deficiency symptoms appear more rapidly and are more severe. A plant that seems to grow well with 9 or 10 hr of high-intensity light may show nutritional problems when the light is given for 15–16 hr per day. We have seen lettuce seedlings and pinto bean plants become nitrogen deficient in a growth chamber while similar plants in the same soil mixture stayed healthy in the greenhouse. Watering with 1 g of soluble 20-20-20 fertilizer per liter of water three times a week prevented the problem. Needless to say, fertilizer treatments are now begun the first day of transfer to the growth chamber. If we wait

TABLE 4.8 NCSU phytotron nutrient

	Grams per liter of stock solution	Grams total for 180 l.
Solution A		
Magnesium nitrate		
$(Mg(NO_3)_2 \cdot 6\ H_2O)$	65.0	11,575
Calcium nitrate		
$(Ca(NO_3)_2 \cdot 4\ H_2O)$	160.0	28,800
Solution B		
Ammonium nitrate		
(NH_4NO_3)	80.0	14,400
Sequestrene 330 Fe*	25.00	4,500
Potassium phosphate		
(KH_2PO_4) (monobasic)	12.0	2,160
Potassium phosphate		
$(K_2HPO_4) \cdot 3\ H_2O$ (dibasic)	14.0	2,520
Potassium sulfate		
(K_2SO_4)	15.0	2,700
Sodium sulfate		
(Na_2SO_4)	17.0	3,060
Boric acid		
(H_3BO_3)	0.70	126
Molybdic acid		
$(MoO_3 \cdot 2\ H_2O)$	0.005	0.9
Hampene Zn**		
(14.5%)	0.045	8.1
Hampol Mn**		
(9.0%)	0.63	113.0
Hampol Cu**		
(9.0%)	0.03	5.4
Sequestrene Co*		
(14.0%)	0.001	0.18

Notes. Use care not to overstir, as excessive agitation may cause precipitation. The minor elements are dissolved before adding to the B stock solution; the final pH should be 7.0. Both stock solutions A and B are to be used at 1:500, or 200 ml stock per 100 1. water.
* Geigy Agricultural Chemicals, Ardsley, New York 10502.
** Hampshire Chemicals, W. R. Grace Co., Nashua, New Hampshire 03060.

several days before fertilizing, the deficiency has already occurred and the plants improve slowly, if at all. Since additional nutrients have to be supplied so frequently to soil culture, we have essentially established a nutrient-solution culture in a soil substrate. Gravel or sand plus a complete nutrient solution should and does work well in growth chambers and prevents many of the problems that arise with a soil substrate. Usually plants have to be watered twice daily, and we find that a 1/4-strength nutrient solution with 1/2-strength nitrogen (table 4.8) applied at every watering period gives satisfactory results for most plants. This nutrient solution, worked out independently by G. M. Jividen, shows a remarkable similarity to the one used at the phytotron in France (Nitsch 1972).

The appearance of nutritional problems in soil that would be considered well fertilized in greenhouses is considerably less frequent if tin cans or plastic pots are used rather than clay ones. Although no detailed studies are available, these nutritional disorders may be related to water stress or to root temperatures (Krizek, Bailey, and Klueter 1971).

Substrate

Satisfactory plants can be grown in a wide variety of substrates if the nutrient solution and water regimes are designed properly. The phytotron in France uses glass wool, and the one at Duke a gravel-vermiculite substrate similar to the mixture used in the original phytotron at the California Institute of Technology. Many researchers prefer river-bottom sand, and Raper (1971) has used it effectively in controlled-environment studies with tobacco.

Plant-growth tests in the NCSU phytotron showed that a mixture of peat moss and vermiculite was superior to sand, gravel, vermiculite, perlite, or any mixtures thereof when watered twice daily in controlled-environment rooms. Figure 4.2 illustrates the response of one of the 15 species tested. Commercial blends of the material, originally called Cornell Mix, are marketed under names such as Jiffy-Mix and Redi-earth. Peat-vermiculite mixtures include nutrient chemicals similar to those added to the Cornell Mix and consequently may be "hot" for some kinds of plants. For example, when they are used as a seed bed for Impatiens or Salvia, deionized water must be used until the seedlings are large enough for transplanting. Weak nutrient solution or even tap water may cause

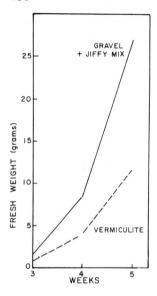

FIG. 4.2. Effect of substrate on growth of First Lady marigold plants. Jiffy-Mix is a commercial adaptation of the peat moss-vermiculite mixture developed at Cornell.

poor growth or complete loss of the seedlings, with symptoms that might be casually treated as damping-off.

Pot Size

Inadequate pot size can curtail growth before the root mass appears to be excessive. Some species like tobacco are exceptionally sensi-

TABLE 4.9 Effect of pot size and temperature on growth of NC-5 peanut after 4 weeks

Day/night temperatures (C)	Pot size (cm)	Fresh weight (g)	Dry weight (g)
30/26	11.4	68	9.54
	15.2	111	13.46
26/22	11.4	35	5.84
	15.2	46	6.43
22/18	11.4	15	3.32
	15.2	18	3.32
18/14	11.4	4	0.53
	15.2	4	0.66

tive to pot-size effects, and 25-cm pots seem to be the minimum for normal growth after transplantation. Others, like soybean and peanut, are less sensitive to root restriction, and the effect of inadequate pot size might escape notice. Peanut plant growth (table 4.9) is significantly lessened by small pots when the temperature is conducive to rapid growth. At temperatures of 22/18 C or lower, plant growth is slowed to the point that pot-size effects are not apparent after 4 weeks' growth. Above 22/18 C, however, pot size begins to restrict growth between the third and fourth weeks after planting.

References

Arthur, J. M., and J. D. Guthrie. 1928. Some effects of light and carbon dioxide on the growth and flowering of plants. *Phys. Therap.* 44:130–36.

Bailey, W. A., D. T. Krizek, and H. H. Klueter. 1969. Controlled environments for agricultural production. In *Controlled Environments Food Technology*, pp. 27–43. Proceedings of a symposium, Winrock Farms, Morrilton, Ark.

Biamonte, R. L. 1972. The effects of light intensity on the initiation and development of flower primordia and growth of geraniums. Master's thesis, North Carolina State University.

Biological Handbook: Biological Data Book. 1964. Washington, D.C.: Federation of American Societies of Experimental Biology.

Biological Handbook: Environmental Biology. 1966. Bethesda, Md.: Federation of American Societies of Experimental Biology.

Biological Handbook: Growth. 1962. Washington, D.C.: Federation of American Societies of Experimental Biology.

Bhatti, D. S. 1971. The Lespedeza cyst nematode, *Heterodera lespedezae,* its life cycle and the influence of certain environmental factors on development and pathogenicity. Ph.D. diss., North Carolina State University.

Borthwick, H. A., and S. B. Hendricks. 1960. Photoperiodism in plants. *Science* 132:1223–28.

Downs, R. J., H. A. Borthwick, and A. A. Piringer. 1958. Comparison of incandescent and fluorescent lamps for lengthening photoperiods. *Proc. Amer. Soc. Hort. Sci.* 71:568–78.

Downs, R. J., S. B. Hendricks, and H. A. Borthwick. 1957. Photoreversible control of elongation of pinto beans and other plants under normal conditions of growth. *Bot. Gaz.* 118:199–208.

Downs, R. J., and A. A. Piringer. 1958. Effects of photoperiod and kinds of supplemental light on vegetative growth of pines. *Forest Sci.* 4:185–95.

Downs, R. J., A. A. Piringer, and G. A. Wiebe. 1959. Effects of photoperiod and kind of supplemental light on growth and reproduction of several varieties of wheat and barley. *Bot. Gaz.* 120:170–77.

Funke, G. L. 1931. On the influence of light of different wavelengths on the growth of plants. *Rec. Trav. Bot. Neerl.* 28:431–85.

Gaastra, P. 1963. Climatic control of photosynthesis and respiration. In *Environmental Control of Plant Growth,* ed. L. T. Evans, pp. 113–40. New York: Academic Press.

Gaastra, P. 1964. Some comparisons between radiation in growth rooms and radiation under natural conditions. *Phytotronique* 1:45–53. Paris: Editions du Centre National de la Recherche Scientifique.

Gardner, R. 1966. Effects of carbon dioxide enrichment on crops of tomato, lettuce and chrysanthemum as grown commercially on the NAAS Experimental Horticultural Stations in England and Wales. In *Proceedings of the XVII International Horticultural Congress,* vol. 1, ed. R. E. Marshall, p. 347. College Park: Univ. of Maryland.

Hellmers, H. 1963. Some temperature and light effects in the growth of Jeffrey pine seedlings. *Forest Sci.* 9:189–201.

Hellmers, H. 1966. Growth response of redwood seedlings to thermoperiodism. *Forest Sci.* 12:276–83.

Hesketh, J. D. 1963. Limitations to photosynthesis responsible for differences among species. *Crop Sci.* 3:493–96.

Hoffman, G. 1971. Humidity effects on yield and water relations of nine crops. Paper no. 71–937, American Society of Agricultural Engineers, St. Joseph, Mich.

Hopen, H. J., and S. K. Ries, 1962. The mutually compensating effect of carbon dioxide concentrations and light intensities on the growth of *Cucumis sativis.* *Proc. Amer. Soc. Hort. Sci.* 81:358–64.

Jividen, G. M., R. J. Downs, and W. T. Smith. 1970. Plant growth under high intensity discharge lamps. Paper no. 70–824, American Society of Agricultural Engineers, St. Joseph, Mich.

Klueter, H. H., W. A. Bailey, P. N. Bolton, and D. T. Krizek. 1971. Xenon light and temperature effects on photosynthesis in cucumbers. Paper no. 71–935, American Society of Agricultural Engineers, St. Joseph, Mich.

Kramer, P. J. 1957. Some effects of various combinations of day and night temperatures and photoperiod on height growth of loblolly pine seedlings. *Forest Sci.* 3:45–55.

Kretchman, D. W., and F. S. Howlett. 1970. CO_2 enrichment for vegetable production. *Trans. ASAE* 13:252–56.

Krizek, D. T., W. A. Bailey, and H. H. Klueter. 1971. Effects of relative humidity and type of container on the growth of F_1 hybrid annuals in controlled environments. *Amer. J. Bot.* 58:544–51.

Madsen, E. 1968. Effect of CO_2 concentration on the accumulation of starch and sugar in tomato leaves. *Plant Physiol.* 21:168–75.

Matsui, T., and H. Eguchi. 1972. Regulation of plant growth and differentiation by automatic program control of environmental factors by use of a computer. I. Regulation of hypocotyl elongation of cucumber by automatic program control of temperature. *Environ. Contr. Biol.* 10:21–27.

Meijer, G. 1959. The spectral dependence of flowering and elongation. *Acta Bot. Neer.* 8:189–246.

Mohr, H. 1972. *Lectures on Photomorphogenesis.* Berlin: Springer-Verlag.

Nitsch, J. P. 1972. Phytotrons: Past achievements and future needs. In *Crop Processes in Controlled Environments,* ed. A. R. Rees, K. E. Cockshull, D. W. Hand, and R. G. Hurd, pp. 33–56. London: Academic Press.

Pallas, J. E., A. R. Bertrand, D. G. Harris, C. B. Elkins, and C. L. Parks. 1965. *Research in Plant Transpiration.* Production Research Report no. 87. Washington, D.C.: U.S. Department of Agriculture.

Parker, M. W., and H. A. Borthwick. 1949. Growth and composition of Biloxi soybean grown in controlled environment with radiation from different carbon-arc sources. *Plant Physiol.* 24:345–58.

Piringer, A. A., and H. M. Cathey. 1960. Effect of photoperiod, kind of supplemental light and temperature on the growth and flowering of petunia plants. *Proc. Amer. Soc. Hort. Sci.* 76:649–60.

Raper, C. D. 1971. Factors affecting the development of flue-cured tobacco grown in artificial environments. III. Morphological behavior of leaves in simulated temperature, light duration and nutritional progressions during growth. *Agron. J.* 63:848–52.

Raper, C. D., and R. J. Downs. 1973. Factors affecting the development of flue-cured tobacco in artificial environments. IV. Effects of carbon dioxide depletion and light intensity. *Agron. J.* 65:247–52.

Van Soest, W. 1966. Practical experience with CO_2 on vegetable growing under glass. *Proceedings of the XVII International Horticultural Congress,* vol. 1, ed. R. E. Marshall, College Park: Univ. of Maryland.

Vince, D., and R. H. Stoughton. 1957. Artificial light in plant experimental work. In *Control of the Plant Environment,* ed. J. P. Hudson pp. 72–82. London: Butterworth.

Wassink, E. C., and C. Van der Scheer. 1950. On the study of the effects of light of various spectral regions on plant growth and development. *Proc. Koninkl. Ned. Akad. Wetenschap.* 53:1065–72.

Went, F. W. 1957. *Environmental Control of Plant Growth.* Chronica Botanica, vol. 17. New York: Ronald Press.

Whittingham, C. P. 1973. The relationship between crop physiology and analytical plant physiology. In *Crop Processes in Controlled Environments,* ed. A. R. Rees, K. E. Cockshull, D. W. Hand, and D. W. Hurd, pp. 177–83. London: Academic Press.

Wittwer, S. H. 1966. Carbon dioxide and its role in plant growth. *Proceedings of the XVII International Horticultural Congress,* vol. 3, ed. H. B. Tukey, pp. 311-22. College Park: Univ. of Maryland.

Wittwer, S. H., and W. Robb. 1964. Carbon dioxide enrichment of greenhouse atmospheres for food crop production. *Econ. Bot.* 18:34–56.

Wiebe, H. J. and H. Krug. 1973. Physiological problems of experiments in growth chambers. Paper presented at the International Society for Horticultural Science Symposium, Hanover, Germany, 1972. *Phytotronic Newsletter* nos. 4–6, p. 6. ed. by P. Chouard and N. de Bilderling, Gif-sur-Yvette, France.

5 SPECIFICATIONS FOR THE PLANT-GROWTH CHAMBER

Except for large installations such as phytotrons, plant-growth chambers are usually purchased by individual biological scientists as tools for specific research objectives. It is not surprising, therefore, that controlled-environment rooms are selected on the basis of performance in the same manner as amino acid analyzers, spectrophotometers, and other research equipment. Unfortunately, the methods by which the contractor meets the performance requirements, along with key maintenance features, are rarely considered at the time of purchase.

The fallacy of selecting a growth chamber by performance alone usually occurs to the owner when he discovers that peak operation is short lived and that breakdowns, with loss of research time and material, are frequent. The owner may suddenly become aware that he has no idea of the preventive maintenance to be performed, that parts lists are not available, and that essential components are neither accessible nor easily repaired. He rightly concludes that the mechanical problems of the controlled-environment room could be significantly reduced by more capable engineering, use of quality components, and improved construction methods. He might in retrospect also conclude that better specifications would have increased the odds of obtaining satisfactory equipment.

Specifications must be written as much for future maintenance and for reliable operation as for performance. The contractor or manufacturer does not have to keep the equipment operating, and few commercial chamber manufacturers have (or seem to want) maintenance histories of their various models. Even when dissatisfied biologists have pointed out the problem areas, there has been no evident rush to make improvements. Ensuring more reliable operation is becoming a major issue in controlled-environment work. Obtaining that reliability is the sole responsibility of the owner, who must write more detailed specifications than the performance data given in the manufacturers' brochures. Since the research scientist is not usually a growth-chamber expert, the AIBS bioinstrumentation council has prepared a set of guidelines to aid the writing of specifications for controlled-environment rooms. Unfortunately, guidelines do not explain the choices involved or supply background material on which to base the specifications.

Theoretically the specifications are part of a legally binding contract, which suggests strongly that they be given great and careful consideration. Yet, inexplicably, plant-growth-chamber specifications are often written in general, ambiguous terms, apparently with the naive expectation that the manufacturer will strive mightily to provide a facility that meets all the owner's requirements. Naturally, most growth-chamber manufacturers and contractors would like to make a superior product, but they are in business to make money and must compete with the less scrupulous. In any field there are always a few suppliers whose only interest is to meet the letter of the specifications at least cost and most profit. Unfortunately this type of contractor often submits the lowest bid because he has diligently searched the specifications for loopholes. The specifications must therefore be written with the unscrupulous contractor in mind.

A very important point to remember is that the intent of what is written in specifications has little value and is certainly unenforceable. The owner must expect from the outset to receive no more than is literally and unmistakably stated in the document. A second important point is to remember that one cannot specify both the performance and the size of equipment used to obtain that performance. One can, however, describe the type of equipment desired, in order to ensure quality components, and one can insist on certain minimum ratings and sizing.

Typically, growth-chamber specifications must consider the following areas:

Dimensions

One interior dimension of the plant-growth room is usually determined by the length of fluorescent lamps. Reach-in cabinets are usually 1.22 m in length, but the width may be 0.61, 0.91, 1.22, or 1.82 m. Medium and large walk-in rooms are usually 2.44 m in one dimension, with the other dimension anywhere from 1.22 to 3.66 m or more. Height of reach-in cabinets is rarely more than 1.22 m, and for walk-in rooms about 2.13 m. Internal dimensions are of interest, however, only in terms of usable space. Thus, one might discover that plant trucks will not fit into a room 1.22 × 2.44 m because of a pipe chase extending into the room near the floor. One might find that a reach-in cabinet has less than the anticipated 0.91×1.22 m of usable area because the housing of an aspirated sensor creates a strong air stream and partially shades plants placed under or near it. The anticipated usable height may be reduced by fans near the top of the walls that not only create undesirable air velocities but actually pull plant leaves into the blades. Thus, a clear distinction must always be made between the maximum internal dimensions and the usable ones.

Outside dimensions are of lesser interest, provided the chamber will fit into the designated space and pass through the doors and halls leading to that space. The stated length and width of prefabricated rooms may not include protruberances like door handles, knobs, and switches. Overall height can be a great source of difficulty because fans, air exhausts, or inlets with filters may be located on top and serviceable only from above. For example, a certain manufacturer makes a room advertised as 2.26 m high, and one might assume that after passing through the doors it would fit under a 2.44-m ceiling. Unfortunately, fans and louvers are on top and the lamp loft is hinged so that it can be raised for lamp replacement—thus requiring at least 0.61 m, and preferably 0.91 m, of additional height.

Floor

A number of commercial growth chambers use the floor as part of the air-flow system or have air ducts below the plant-growing bench. Floors are not optional in these types of chambers. When the floor is part of the chamber, a drain must be included to remove water spillage. The drain should be large enough (at least 4 cm) that it is not quickly plugged by pot labels, peat moss, soil, and other substrate material inevitably washed from the pots. A method of cleaning the plugged drain should also be considered; if that method requires removal of plant material from the chambers, another way should be found.

Rooms with downward air flow can often be designed without a floor and thus are readily adaptable to plant-trucking operations. Placing such a chamber over a floor drain on grade will probably not present problems. Placing the plant-growth chamber on a higher story, above grade, may present heat-transfer problems that result in sweating of the underside of the floor, i.e., the ceiling of the story below. If the chambers are to be part of a new building, proper insulation, waterproofing, and

TABLE 5.1 Loading from various sizes of plastic pots filled with river-bottom sand and watered to field capacity

Diameter		Volume		Filled weight		Loading [a]		No. of
(cm)	(in.)	(l.)	(cu ft)	(kg)	(lb)	(kg/m²)	(lb/ft²)	pots [a]
11.4	4½	0.60	0.021	0.95	2.1	78.1	16.0	90
15.2	6	1.57	0.055	2.49	5.5	107.4	22.0	40
20.3	8	3.65	0.129	5.85	12.9	146.9	30.0	28
25.4	10	8.30	0.293	13.29	29.3	190.6	39.0	16

[a] Based on a 3 × 4-ft room.

floor drains at the chamber location can be included in the building design.

Whether floors, benches, or shelves are used, some indication of maximal loading is often necessary. One might, for example, specify 244 kg/m², although to our knowledge no one ever determines if such specifications are actually met, and few manufacturers ever test the designed loading capacity of their chambers. One can, however, estimate the actual loading that might occur. A 15-cm plastic pot that is filled with 2/3 sandy loam and 1/3 peat and watered to field capacity weighs about 2.5 kg. Since four pots will cover about 0.09 m² of growing space, loading would be 107 kg/m². A 25.4-cm pot holds the equivalent of five 15-cm pots, and four 25.4-cm pots cover 0.25 m². Loading with 25.4-cm pots would therefore amount to over 190 kg/m² (table 5.1). Large flats or soil beds used in population-dynamics studies can exert loads of 488 kg/m².

If the plant bed or shelf is to be adjustable, a description of the adjusting system is necessary. The most pertinent factors are whether the adjustment is to be manual or motorized, and whether it will be from inside or outside the growing space. Since a bench-adjusting system that rusts soon becomes useless, specifications should state the material required or obtain a complete replacement warranty. For example, a nonferrous bench may have a steel adjusting mechanism that because of rust becomes completely inoperable after only a few months use.

Walls

The type of exterior and interior wall surface should probably be specified. Exterior walls are considered only for appearance but decisions about the interior walls must consider reflectance and maintenance of that reflectance. White paint and specular aluminum are the most commonly used wall surfaces. Both have a high reflectance, reasonably long life, and good maintenance characteristics. The desirability of other types of finishes is highly questionable. Aluminized Mylar, for example, is easily damaged, would probably need considerable maintenance in repairing tears, and has been poorly installed in many cases. Moreover, there are several types of aluminized films that have considerable differences in reflectance, so the best type would have to be specified by name. Materials like stainless steel have such poor reflectance characteristics (55%) that they are rarely considered except in special applications such as seed germinators.

Insulation 5 cm thick is marginal, especially when the interior temperature exceeds the outer temperature by more than 10–14 C. Material 7.6 cm thick costs about 20% more than 5-cm insulation and is well worth the added investment in custom-built installations that will be used at 10 C or lower. Commercial chamber manufacturers would probably increase the price more than 20%, especially for purchase of a relatively small quantity or when manufacturing procedures must be altered.

Side-wall heat exchangers are sometimes designed so that the chamber wall forms three of the sides and a thin, uninsulated panel is used for the side facing the growing area. Since the inlet air is almost always colder than the room air, these thin walls will be cooler than the other walls of the room. Measurements show that with about 270 hlx and a 26-C air temperature, the thin wall is at 24.5 C and the insulated walls are at 27.5 C. Lowering the room temperature to 16 C results in a thin wall at 13.5 C and an insulated wall at 17 C. The cold-sink effect can be avoided by using a 2.5-cm layer of rigid insulation, which also increases the strength of the thin, often poorly braced, panels and reduces flutter, vibration, and noise.

Doors

The entire chamber should be light-tight, including the door area. Light-tight means impermeable to light, but the term is subject to a much less precise interpretation unless accompanied by additional information, such as the way in which the degree of darkness is to be determined. If one enters a controlled-environment space with lights out and remains there 15–20 min to become dark adapted, one can easily detect light leaks. These leaks can be biologically effective, since a few tenths of a lux throughout the dark period can prevent flowering in some kinds of short-day plants (Borthwick and Parker 1938). A good test method might be to expose a sheet of Tri-X film for some appropriate time period, perhaps 30 min. If no increase in density occurs, the room can be considered biologically light tight.

Doors should also be as air-tight as possible. The AIBS guidelines suggest stating the allowable leakage and the inside-outside pressure difference. Unfortunately neither the guidelines nor the literature provides any data from which such specifications might be drawn, or any way to determine if they are met. In most chambers the pressure differential is quite small; even when outside air is introduced, it is exhausted at the same rate. The more air-tight the room, the more easily it could later be adapted for control of atmospheric content.

Door openings in walk-in chambers should be no smaller than 96 × 183 cm, and for reach-in cabinets at least 50 × 75 cm. Growth rooms of intermediate size, 1.22 × 2.44 m, need two doors on the same side if the investigator is to reach all plant material easily. Some chambers with one door have a removable bench section to allow the operator to enter the room far enough to reach the corner plants. Plants would of course have to be removed from the bench section and placed outside in an uncontrolled environment at each watering. As a result, the bench section commonly is removed permanently. Not only is air flow uniformity upset, but also the loss of growing space, 0.18 to 0.28 m² (nearly 10%), can hardly be considered minor or inexpensive.

Doors fastened with magnetic gaskets or magnet and strikers are generally preferred over latched ones. All latching door hardware must, for safety reasons, have an opening device on the inside. The device is difficult to provide without broaching the insulation and setting up a metal thermal conductor to the outside. At least two magnet-strikers must be used with each door if it is to continue to close properly after several months of use. In one type of installation the magnets are glued in place. After a period of use, the glue bond breaks and the magnets fall out. Since reinstalling the magnets requires the door to be open several inches for at least 24 hr, invariably disrupting experiments, this method of door fastening can only be recommended if the specifications include a permanent method of securing the magnet, as well as a system for adjusting the strikers. Magnetic gaskets provide the best method of keeping doors closed and at the same time provide a good seal against air and light leaks.

If the doors are to continue to close properly, at least three regular hinges or a piano hinge should be used on each door. Prefabricated chambers are often installed on floors that are not exactly flat; the resultant warpage of the chamber can easily cause the doors to fit poorly. The fewer and smaller the hinges, the greater the trouble encountered after several months' use.

In many applications the growth rooms use an existing floor, and for convenience of operation no sill is desired. An immediate problem arises in making the door light-tight at the base, and no good solutions have yet been provided. Automatic door bottoms such as those used at the Southeastern Plant Environment Laboratories are not satisfactory because of the large amount of maintenance required to keep them functioning. Compressed-air-inflated seals are used on the chambers at the University of Pennsylvania, and at the Wisconsin biotron the sliding doors drop at the closed point. These systems seem to work satisfactorily, but no data are available on the degree of maintenance required.

Observation ports in doors are useful for inspecting the plants without disturbing the environmental conditions. Port covers must be sturdily made, however, or they will allow light to leak into the room. The best types use a magnetic gasket around the entire opening. One-way glass may help reduce light leaks, but since the material transmits 4–8% of the light it is not recommended unless a cover is provided. A lock should be provided for the port covers to prevent casual opening when the room is dark, thereby upsetting the photoperiod regime.

Penetration ports. Capped or sealed holes in the chamber wall for the possible installation of tubing and wires can rarely be planned in advance. Therefore, unless there is an immediate use for such penetration ports, they should not be included in the specifications. PVC pipe with styrofoam plugs makes very satisfactory penetration ports that can be installed by the owner, when and where needed.

Temperature

Range. A temperature range of 7–37 C is a reasonable and realistic specification that can be met without difficulty. Higher temperatures may result in increased maintenance because of shorter life of many components. Fan motors, for example, may have windings rated at 150 C, meaning the total temperature of the winding; at this temperature almost immediate failure will occur. However, the higher the ambient temperature and the closer the windings get to the design rating, the shorter their life. If growth-room temperatures are going to exceed 37 C for much of the time, then high-temperature components should be specified.

Temperatures below 7 C usually require a coil-defrost system; otherwise ice will form, with subsequent loss of temperature control. Few automatic defrost systems operate without causing an increase in room temperature at every defrost cycle. In one case where the owner simply specified an automatic defrost system, the temperature rose 7–8 C each time the system operated. This particular method heats the secondary coolant and runs it through half of the coils at a time. Theoretically the system should have worked better, so the owner spent many man-hours trying to improve it. At a room temperature of 0 C the best performance obtained was a 2-C rise every 30 min, resulting in a temperature above the set point 25% of the time.

Direct-expansion systems may use a hot gas defrost method, but in a small chamber with only one evaporator the result is usually not satisfactory. From the information in the refrigeration section, it is clear that all the vapor from the condenser could flow to the evaporator and condense there. Then the vapor pressure of the whole system would equalize to the evaporator pressure and flow would stop. With no vapor supplied to the compressor and consequently no heat of compression, defrosting would be slow. In a room with two evaporators, one can be defrosted at a time and the system can work fairly well, especially if air can bypass the defrosting (and consequently hot) evaporator.

Controls. Electronic controls can hold temperature variations at the control point to ± 0.25 C in a well-designed chamber. If air flow distribution and uniformity are poor, the variation can be 3–4.5 C in as small an area as 0.2 m². Spatial variation improves when the lights are off and is almost always less in an empty chamber. Specifications should state allowable spatial variation in the standard 1-m plane of reference, as well as the light conditions under which chamber temperature performance is to be evaluated.

Simply describing the ability of the controls is inadequate—for example, quoting from specifications written for controlled-environment facilities by engineers who normally do air conditioning of office buildings. The controls were to be "capable of operating over a temperature range of −18 to 48 C with a maximum control differential of ±0.4 C with a resistance or thermocouple type sensor." The result of these specifications was ±2 C variation about the set point and a spatial variation of 6–8 C. In addition, the set point depended on the amount of light available because the sensor was improperly shielded. To the engineer's specifications must be added the exact performance demanded or a detailed account of sensor location, shielding, and aspiration.

Thermal load. The AIBS guidelines (Busser 1971) suggest that data on thermal load be included. Such data are difficult to estimate, but the guideline suggestions are probably satisfactory. They suggest 4 man-hours per day spent inside the chamber, with the door open for half of each hour into a surrounding space at 25 C. They also suggest using the maximum pot load at one time, with the material at 35 C. In most facilities these thermal load values are undoubtedly on the high side. The poor practice of leaving the door open for half-hour periods needlessly introduces considerable temperature variation and it surely cannot be a common practice.

Air Flow

Optimum air velocity seems to be accepted as 30 m/min for most plant-growth-chamber applications, although some investigators (Van Bavel 1970) have advocated rates 10-fold greater. Adjustable air flow is possible, but performance may worsen at air flows above or below the design conditions of the coil. Coil ratings are based on a design face velocity, and when air flow is reduced the performance is affected. At high velocities, water condensed on the evaporator may carry over into the growing area. Whatever velocity is desired should be clearly stated as mean velocity, with permissible maxima and minima as well as duration of the extremes, with a carefully described plant and pot load. If more than one air velocity is required, the permissible temperature should be stated at each flow rate. Controlled-environment rooms with programmable air velocity can be built only when consideration is given to the effect of air flow on the design parameters.

Direction of air flow should be stated. Although I believe that downward, top-to-bottom flow is best both physically and biologically, many apparently satisfactory chambers have been built using bottom-to-top flow.

With small seedlings and with plants widely spaced, current methods of air distribution seem to result in satisfactory plant growth. Large plants that are close enough together to restrict air flow may benefit from auxiliary circulating fans to improve gas exchange between leaves and air.

Relative Humidity

As long as the range of water vapor can be delineated on a psychrometric chart, the likelihood of unrealistic demand accompanied by high cost is considerably reduced. The psychrometric chart can also keep the requirement for percentage of variation within reasonable limits. Limiting factors discussed earlier, such as coil temperature and the ability of various kinds of devices to put water into the air or to remove it, must be considered at the time the specifications are written. Do not ask for greater humidity control than is really necessary or for precision greater than the plant's ability to detect.

Carbon Dioxide Control

In the future, CO_2 control will be given much more attention than it has received in the past. Owing to general lack of experience in this area by plant-growth-chamber contractors and manufacturers, it would probably be more efficient for the owner to add the CO_2 control after the chambers have been accepted. Commercial chambers are sometimes advertised as having a CO_2 controller but this is very misleading since they usually provide only a flow meter and metering valve. As was pointed out previously, makeup air systems are largely unsatisfactory. Some arrangement therefore must be planned for CO_2 injection and of course measurement. To aid in the installation of a CO_2 system, plant-growth-chamber specifications could include solenoid valves actuated by the lighting system, flow meters, and entry ports for the CO_2 supply. Most important is ensuring that the chamber will be sufficiently air-tight to allow enhancement of CO_2 as well as control at ambient levels. W. A. Bailey (personal communication, 1973) has recommended that the rate of decay should not exceed 10 ppm per minute. Tests can be made by filling the room with CO_2 to a level of 1000 ppm. Measurements begin when the CO_2 concentration inside the space is 500 ppm higher than outside. If the ambient concentration is 350 ppm, records of CO_2 level would begin at 850 ppm and should require at least 50 min to reach the ambient level.

Illumination

The fluorescent-incandescent system. The quantity and quality of the light are best specified by naming the types of light sources to be used, along with the number and wattage of each kind of lamps. Spatial uniformity specifications should be made for each light source, separately as well as together, as a function of percentage variation from the intensity in the center of the room at 1 m from the light source or barrier.

Very few investigators know how much energy above 800 nm or below 400 nm is required or desirable. Many of us even have difficulty in relating plant growth to total energy or to energy between 400 and 700 nm, especially when energy is described in the literature in terms of a wide variety of units such as microwatts per square centimeter, Langleys, kilogram-calories, joules, and ergs per square centimeter per second. Usually the investigator can relate illumination to plant growth. He knows, for example, that with a reasonably prudent lamp-changing schedule a level of 430 hlx can be maintained in a 1.22×2.44-m room with 28 fluorescent lamps of 215 w and 16 incandescent lamps of 100 w each. Almost every growth-chamber owner can state the illumination level in his room but not one in twenty can say how much energy, total or between 400 and 700 nm, he has at plant level. The availability of the PAR meter is certain to induce biologists to use incident quanta between 400 and 700 nm as a growth-chamber specification, especially since $\mu E\ m^{-2}\ sec^{-1}$ can easily be related to a personal concept of light intensity.

Attempts to state illuminance or energy levels as a specification will lead to considerable unhappiness unless accompanied by greater detail. For example, the measurement distance from the lamps or from the barrier should be stated as 1 meter. The time of measurement in terms of hours of lamp use and number of starts must be included, because a room with an illuminance of 645 hlx after 100 hr of operation can easily be below 430 hlx after a few months. If the lamp system is ventilated with outside air, the light level may vary with the season. The measuring device must be fully described, and if the manufacturer's or contractor's instrument is to be used, a recent certificate of calibration would seem to be a reasonable requirement. If the owner's instrument is to be used, the data could be challenged unless it too had been recently calibrated.

Control of light intensity. As previously discussed, the highly loaded fluorescent lamps used in plant-growth chambers are not readily amenable to dimming. It can be done, of course, but the cost of dimming alone may well equal the total cost of the chamber and cannot be considered economically realistic. Step control of intensity in 1/3, 2/3, and 3/3 of the lamps is inadequate, in my opinion, and I would recommend a minimum of six or seven steps of intensity.

Barrier. Many controlled-environment rooms use a transparent barrier between the lamps and the growing area. This barrier may have to be removed for lamp replacement, and requests such as "easy removal" are completely inadequate. Barrier removal may require two men and the room one-third to half empty of plants, but if the contractor claims that the system is "easy" then it meets the specification. Considering that the barrier needs to be removed at least six times a year, the lack of improved designs is surprising. For instance, the barrier could be hinged in sections small enough to be lowered by a single person, usually without removing any plants, or one section could be made to slide over another. Most barriers are made of rigid plastic, and the quality and type should be specified. Many plastic barriers are too thin, and warp and sag excessively. The minimum thickness for plastic barriers should be 0.5 cm if the sheets are no larger than 1.22×1.22 m and are supported on all sides.

Lamp cooling. The method of cooling the lamps should be fully described. Detailed instructions should be given on access to fans, lamp sockets, and any other equipment located in the lamp loft. If the lamp loft is to be air conditioned, the temperature limits and uniformity should be clearly stated. Although the manufacturers have the responsibility of sizing the heat exchanger to provide the specified temperature conditions, the owner must ensure that it can be serviced. Water or secondary-coolant coils will need to be reverse flushed and entrained air will need to be removed. In at least one type of lamp loft design used by growth-chamber manufacturers to meet most specifications, bleeding the air from the coils requires removal of the barrier, removal of all lamps, removal of 1.22 × 1.22-m sheet-metal panels, and the use of an offset screwdriver for blind operation of the bleeder valve. Moreover, this particular system has no provision for flushing. Satisfactory operation will require a major redesign at the expense of the owner. It should not be necessary to cut into pipe or dismantle the lamp loft to perform common maintenance tasks, but only through proper writing of the specifications can the owner ensure against such design faults.

The direction of air flow in the lamp loft depends on the type of lamp being used and should be specified. It seems to be much easier to construct air-conditioned lamp lofts with air flow for the Power Groove lamp than for the T-12, and without specifications the wrong type of air flow is likely to be used. Ventilated lamp-cooling systems are especially likely to be designed with the air-flow direction suitable for Power Groove lamps. The lamp air-flow system at the Wisconsin biotron provides a good model of effective design. In this case, cooling air is injected through slots running directly over the entire length of the lamps. Modifications might include restricting portions of the inlet slot to increase air flow over the critical parts of the lamps.

The minimum linear air flow should also be specified, especially in ventilated systems. Until better data are available, air velocities must be based on lamp-company information, which admittedly is obtained under conditions that do not resemble those in the plant-growth-chamber loft. These data indicate, however, that air velocities for lamp cooling should be at least 50 m/min (fig. 2.16).

Lamp-loft ventilating or cooling air should pass between the lamps and the barrier, either rising through the lamps to return and discharge or dropping between and over them to reach the discharge grills. Directing the air flow against the upper, usually hot, side of the barrier will reduce heat conduction through it, and forcing the air between the lamps instead of just across the upper side will improve cooling conditions and consequently light intensity.

Ballasts. Methods of mounting ballasts often fail to provide sufficient cooling. Ballasts in poorly cooled cabinets may not fail for 2 or 3 years, but the failure is still premature. Reduction of this maintenance problem requires specifications that describe exactly how the ballasts are to be cooled or state the maximum permissible temperature. The system should be tested and temperature rises measured before acceptance. As an example, in one installation using the Vacha lighting system, the cabinets (not designed by Vacha) run too hot. Premature ballast failure was predicted as soon as the owner saw the cabinet layout because, among other problems, a large transformer was mounted just below the inlet vent. Considering the dire predictions of the owner, the architect informed the contractor that he was "responsible for proper cooling which will be demonstrated by satisfactory long-term performance." Satisfactory long-term performance has not been obtained, however, and, as most growth-chamber owners soon discover, a contractor rarely accepts any responsibility for long-term performance. During summer months the transformer raises the inlet air to a temperature exceeding the permissible ambient conditions, 40 C, for the chokes. If the specifications had stated that the air entering the ballast cabinet could not exceed the temperature of the air at 1 m from the cabinet wall and 1 m from the floor, the system might have worked, especially if a fan ventilation system had been included.

The only way to ensure proper ballast temperature is by forced draft ventilation. Although few data are available, air passing the ballasts at 90 m/min seems to provide satisfactory cooling. Commercial ballasts should be mounted vertically, about 5 cm apart, on bars or spacers that keep them 2–3 cm from the wall. As the air passes through the ballast cabinet, however, it increases in temperature; if it passes more than five ballasts the heat gain could be excessive.

Safety Features

Off-normal alarms and limit switches should be an integral part of all controlled-environment facilities. When the temperature exceeds the set point, all lights and heaters should be switched off. This requirement should be stated in detail, with all items listed. It may seem obvious that the heaters as well as the lamps should be included, especially in pre-fabricated units when the manufacturer has switched off both lamps and heaters in earlier systems. If switching off of heaters is not specified, however, it may not be included, even though it only requires one wire about 6 m long and an insignificant amount of labor at the time of construction. Modification after construction requires the same 6 m of wire, but the chamber will have to be shut down and dismantled in order to pull the wire from the front to the rear, requiring 6–8 man-hours of labor.

Protection from freezing should also be specified; the most straightforward method is simply to shut down the coolant supply. Closing of the coolant valve on low-temperature alarm is not satisfactory by itself, because the low, off-normal condition may have resulted from the valve's sticking open and consequent inability to respond to the control signal. A separate, alarm-operated shut-off valve could be installed at reasonable cost or, where used, the pump could be turned off.

These safety devices should all be tested prior to acceptance of the equipment to ensure that the specifications are met and all thermal switches are calibrated. Some contractors install the thermostats without any calibration. It pays to remember that modifications after chambers are assembled usually require many more man-hours than those made during assembly. The low-temperature safety features are a good illustration; they would require 1 man-day per chamber to install after ownership, compared to only a few minutes additional time during assembly.

Access to components. Never use the term "easy access" when referring to components requiring maintenance. Never simply require "access panels" to such components. Such vague specifications can result in impossible construction that requires shutting down the complete system to perform what should have been a routine task.

Access panels should of course be provided for all fans, controls, and electrical components. More specifically, the access panels should be hinged and secured with quick-release fasteners. All secondary-coolant coils will need to be reverse flushed, and a valved facility should be provided for this task. All water or coolant coils will need to have entrained air removed. The bleed-off valves should be located so that no dismantling of the chamber is necessary in order to reach them. The bleed-off valves should either protrude from the coil housing or at most be covered by a small panel held by about four screws. All pumps and valves on secondary-coolant or water systems should be installed with flanges or unions so they can be removed without cutting into the pipe. Do not use the phrases "where necessary" or "wherever required to permit removal and maintenance of equipment, valves, etc." If drawings are made, insist on showing all flanges and unions. If only written specifications are used, clearly state every case for which such devices are necessary and also make a general statement concerning all such equipment.

All components of the controlled-environment room should be clearly labeled as to function. Relays, switches, and balancing potentiometers should be identified, as should fuses and circuit breakers. All wiring should be color-coded and numbered at each terminal. Lamp holders should be related to the ballasts that operate them and, if three-phase electrical supply is used, the lamp-phase identification should be included. All piping should be labeled, and the direction of flow indicated. Where labels are not feasible, a durable schematic drawing should be permanently attached to each access area. A complete set of "as-built" drawings should be required with each type of controlled-environment facility.

Parts list. A complete list of all parts, by manufacturer, serial number, and size, should be obtained for every controlled-environment chamber, whether manufactured or contracted. Such a list will save the owner considerable time when various parts have to be replaced, especially if a vendor and price can be obtained for each item in advance.

The Refrigeration System

Strainers and driers should be required in the liquid line, along with a sight-glass liquid indicator installed so that it is clearly visible. The drier and strainer must be accessible in such a way that they can be removed without dismantling anything else. All Freon piping should be silver soldered, and the pipe should be nitrogen-filled during the soldering process to prevent internal scale formation.

The compressor should be vibrationally isolated from the growth-chamber frame. In far too many plant-growth chambers, vibration from the compressor is translated to the growing area. Spring-type vibration isolators should be used. The little rubber feet used on many of the smaller compressors have not proven satisfactory.

An effort should be made to ensure that the condenser unit is large enough. Smaller condensers are cheaper and may thus be installed by a low bidder. Operation may then be marginal, and minimum design temperatures may not be attainable during hot summer months. Power requirements will surely increase since the discharge temperature of the condenser will be higher. Small, water-cooled condensers may have to be reverse flushed each week in areas where water towers collect debris such as oak catkins, pine pollen, soot, and dust. A prefilter or trapping system could be used but is usually omitted because of a false sense of economy that reduces initial cost while raising operating expenses. In larger systems the water-cooled condenser—usually a tube and shell type—will need to be rodded out periodically. Make sure that sufficient unimpeded space, about 1.5 times the condenser length, is provided for this operation.

References

Borthwick, H. A., and M. W. Parker. 1938. Photoperiodic perception in Biloxi soybeans. *Bot. Gaz.* 100:274–87.

Busser, J. H. 1971. *Controlled Environment Enclosure Guidelines.* BIAC Information Module M21. Washington, D.C.: American Institute of Biological Scientists.

Van Bavel, C. H. M. 1970. Towards realistic simulation of the natural plant climate. Paper presented at the UNESCO Symposium on Plant Response to Climatic Factors, Uppsala, Sweden.

6 TESTING AND MAINTENANCE OF THE PLANT-GROWTH CHAMBER

A number of years ago W. Kalbfleish (ASHRAE 1964) recommended a set of standard techniques for measurement of growth-chamber performance. We see no reason to alter his procedures except to use the metric system. In rooms that are 1.22 x 2.44 m or larger, all measurements (including light, temperature, RH, and air velocity) are to be taken in two planes of reference, at 1 m and at 60 cm from the barrier or from the light source when a barrier is not used. Measurements are to be taken at the intersections of a 30-cm grid in these planes. Vertical gradients would be measured on the same grid intersects. Isolines of temperature, wind, and light could then be plotted to describe the conditions of the facility. In rooms smaller than 1.22 x 2.44 m, measurements would also be taken on the 30-cm grid but at 30 and 60 cm from the barrier.

Temperature may be difficult to measure in high-radiant-energy situations because of the influence of the mass and absorptivity of the sensor. For example, a thermocouple no larger than No. 24 wire, with a welded and polished junction 0.5 mm in diameter, still picks up some radiant energy. Nonetheless, this type of sensor is the best overall compromise for determinations of uniformity.

Light uniformity can be established in relative terms as percentage of variation, and a cosine-corrected light meter would be adequate. Light variations should not exceed 5% over the plant-growing area. Intensity, which might be specified in terms of photosynthetically active radiation (PAR), should be more than 700 μE m^{-2} sec^{-1} in a modern fluorescent-incandescent–lighted room after 1000 hr of operation. If it is not, then lamp-cooling methods are suspect.

When the controlled-environment room is ready for acceptance and supposedly running properly, a set of normalcy data should be obtained and filed for future reference. These data would include the original measurements of environmental uniformity plus the operating temperatures of the lamps and ballasts. In addition, voltage and current characteristics of all electrical components, face velocity of all diffusers or discharge velocity of fans, and pressure drop across all pumps and coils should be recorded. Normal operating data for compressors, such as amperage and head pressure, should also be recorded.

Warranty

In most cases the warranty only replaces defective parts—by mail. The owner must troubleshoot the equipment and locate the problem. The owner must remove the defective part and install the new one. Some warranties even require the owner to prove that the part was defective when it left the factory. In other words, the usual warranty is of little or no value to the biologist using the plant-growth chamber.

A warranty should be spelled out in the specifications, to cover whatever the owner believes will be needed. A 1-year warranty in which the manufacturer completely repairs all malfunctions with his own technician or by contract to a local repairman would seem reasonable. The work covered might include more than replacing defective parts or repairing a total breakdown. For example, temperature control may drift from the set point and require frequent recalibration. If the warranty included accuracy of temperature control, the contractor's representative would be required to recalibrate the system whenever necessary. It is essential to include some method of encouraging the contractor to abide by the terms of the warranty. The best method is to withhold a portion of payment. This money can then be used to repair the equipment if the contractor reneges on the contract.

A simple statement of guarantee would be sufficient for a reputable contractor or manufacturer. Since there is usually no real assurance as to who will finally provide the equipment, the warranty, like the specifications, must be written with the unscrupulous contractor in mind.

The cost of a carefully worded warranty is inversely proportional to the reliability of the product. If the chamber has a reasonable degree of reliability, such a warranty should not add much to the cost of the equipment.

Contractor's or Manufacturer's Responsibility

Many of the firms dealing with controlled-environment facilities show little or no interest in the equipment or in the owners' problems after the sale is completed. Most of the manufacturers attempt to meet the letter of the specifications, however, and a few even try to cope with the intent of the document.

If the controlled-environment facility matches the construction and performance requirements of the specifications, the contractor has fulfilled his legal responsibility. If certain components fail prematurely and repeatedly, the contractor has the moral responsibility to go beyond simple replacement with the same kind of unsatisfactory part. He should assist the owner in selecting a better, more reliable component or even redesigning that part of the system since the improvement will be to his benefit in future chambers built by the firm. Some contractors and manufacturers of plant-growth chambers probably accept this moral responsibility, but the majority do not.

The best way to determine whether a contractor for a controlled-environment facility performs responsibly is to ask other owners of the firm's equipment—not the architect and consulting engineer, if they are involved, or the business office, but the person actually using the facility.

Owner's Responsibility

The first responsibility of the owner is to write unambiguous specifications that include every item of equipment and a thorough description of the performance expected. When the equipment is set up and operating, the owner should determine as promptly as possible whether the system meets the specifications as written. He may and probably should make the tests in cooperation with the contractor. The owner abrogates his responsibility when he asks the contractor to provide "certified" performance data, whose correctness the contractor attests to authoritatively, with himself as the authority. Moreover, it seems a naive approach, since the contractor must inevitably be biased.

By drawing upon other users' experience, providing sufficient attention to detail, and writing comprehensive but not excessive specifications, the owner can obtain reliable controlled-environment facilities that will increase his research capability.

Maintenance

Prefabricated controlled-environment rooms are usually constructed in whatever manner is most convenient or economical for the manufacturer. The same statement could be made for rooms that are built in as integral parts of a building. Ease of maintenance and repair seldom receives any real attention in the design, because the manufacturer or the engineer designing the built-in systems does not get involved in the subsequent operation of the equipment. In one prefabricated room, for example, oiling of the blowers requires removal of 8 slot-headed screws, a 30 × 100-cm panel, 28 Phillips-head screws, then the efforts of two men to remove the fan panel and apply a few drops of oil to each fan. This procedure is, of course, followed in reverse to reinstall the system. A little redesign effort showed that use of quick-release fasteners and hinged panels allowed one person to apply the oil at the cost of about 5 min of work.

Replacement of defective parts is equally problematic because many components are installed as if they will never need to be replaced. In one type of reach-in chamber, replacement of light-cap ventilation fans required the hinged light cap to be raised (although no handles were provided), all lamps to be removed, then two men, one outside and one inside the light cap, to remove the hold-down bolts. Power had to be shut off and wires cut for fan removal. The owner designed a simple modification, and the fans are now installed so they can be removed by one man from outside the chamber, without removing any lamps and without loss of power.

Fan maintenance can be so time consuming, if a number of chambers are involved, that it may require a half-time employee to keep the fans all oiled and running. Proper modifications, specified at the time the chambers are purchased, could reduce this problem so that unskilled personnel would need about 10 min to change a fan without any appreciable loss of environmental control. In other words, fan maintenance in a large installation can be reduced from 15 man-days per month to about 1 man-day per month by designing with maintenance in mind.

A preventive maintenance program should be an integral part of controlled-environment-room operation. Some manufacturers of prefabricated plant-growth chambers reportedly have begun to supply maintenance handbooks with their equipment, but I have never seen them and cannot attest to their value.

If no maintenance schedule is provided by the manufacturer or contractor, then one should be set up by the owner. It should be put into

TABLE 6.1 Typical monthly maintenance schedule for a plant-growth chamber

			Environmental check [a]				
Month	Light	Temp-erature, relative humidity	Barrier cleaned [b]	Fans and blowers oiled	Refrig-eration system [c]	Electrical, electronic system [d]	Chamber cleaned [e]
Jan.							
Feb.							
Mar.							
Apr.							
May							
June							
July							
Aug.							
Sept.							
Oct.							
Nov.							
Dec.							

[a] These checks are to be made with instrumentation other than that attached to the chamber. Monitors on the chamber and the control system would be recalibrated and rebalanced whenever the monthly checks showed it to be necessary.

[b] Frequency of cleaning depends on locality and type of light-cap cooling or ventilating system.

[c] Includes cleaning the coil.

[d] Includes checking all relays for burned contacts and all electrical junctions to ensure that hold-down screws and bolts are firmly set.

[e] Includes washing walls and cleaning floor drain. Algicide may be necessary on bench and floor, and the chamber may need to be fumigated.

effect as soon as the chamber is operating. A typical program is illustrated in table 6.1.

A most important part of the maintenance program is a log book for each chamber, in which all maintenance and repair operations are recorded in detail. The use of a log book, rather than memory, will enable the operator to note repetitive breakdowns which indicate a weakness that must be removed. Repairing the same fault again and again is not only irritating but also wasteful of research time and material. Relays that have sticking or burning contacts might be replaced with thyristor (Triac) switches of proper size. Control circuit relays may need to be exchanged for encapsulated units with gold contacts. Premature ballast failure will soon be indicated by the maintenance records and should persuade the owner to install forced ventilation. Repetitive fan failure should lead rapidly to replacement with fans of a more suitable type.

Although troubleshooting and repair are jobs for a mechanic, knowledge of some common sources of malfunction can often allow the investigator to restore a faulty system without loss of time or experimental material. Troubleshooting guides and manuals are not generally available for plant-growth chambers, even for the commercial, prefabricated models. Since manufacturers could have the experienced staffs of large phytotrons prepare a maintenance manual for about the cost of one average growth chamber, the lack of such manuals can only be attributed to indifference.

Troubleshooting guides and maintenance manuals have been prepared by several manufacturers of controls and refrigeration equipment. Copies of the appropriate manuals would prove very useful to the owner of a controlled-environment room (Price 1970).

Refrigeration

Shortage of refrigerant is probably the major source of trouble in a plant-growth chamber that uses a direct-expansion refrigeration system. Leakage is the most common way to lose refrigerant, although the same symptoms can be caused by a faulty expansion valve or a plugged drier. Low suction pressure, frosting of the coil, and bubbles in the sight glass may indicate refrigerant shortage.

If the refrigerant supply falls low enough, the system pressure can drop below atmospheric pressure, and air and moisture can be drawn in. At the rather high operating temperatures of the modern compressor, the air drawn into the system may lead to formation of corrosive products that will cause subsequent mechanical failures. For example, air and high temperatures can cause oil sludging through the formation of organic acids. Organic acids then combine with iron and other metals to form salts that further contribute to the sludge. In the presence of moisture, inorganic acids such as hydrochloric and hydrofluoric form from the refrigerant, which is usually some form of chlorofluoromethane. In fact, water is probably the most detrimental factor in a refrigeration system; levels above 15 ppm are considered to exceed the safe limits for commonly used refrigerants (*Service Pointers* 1960).

Another common malfunction is cycling or shutdown of the compressor because of high head pressure. With air-cooled condensers, a reduced supply of air or excessively hot air would be the most likely problem. Condenser coils of fans may be dirty, fan speeds may be too slow, or the fans may switch off on overload. If the installation is new, the high head pressures may be due to poor site selection. For example, wind or obstructions can cause some of the discharge air to enter the inlet side, thereby raising the temperature. When condensers are installed so that the discharge of one becomes the supply of the other, the air temperatures in the second unit will of course rise to abnormal levels.

With water-cooled condensers, high head pressure will result from any problem that reduces the water flow below designed levels or alters the heat-exchange capacity of the unit. For example, air in the system, plugged screens, or pump failure will reduce water flow. In a new system, especially one using multiple units from a fixed flow source like a cooling tower, improper balancing of flows and marginal design are likely trouble areas. Dirty or scaled condenser tubes reduce heat-exchange capacity. If water flow rates are marginally designed, the tubes may need to be rodded clean every year. The small water-cooled condensers used on the direct-expansion units common to reach-in plant-growth chambers tend to become plugged with silt and debris from the water tower. So common is the problem that reverse flushing of these units ought to be the first step taken whenever the temperatures in these rooms begins to rise as it would with low refrigerant gas.

Refrigerant in the oil reduces viscosity and can cause foaming. In secondary-coolant systems, more than one compresser may be used and an idle one may allow a rise in vapor pressure in the crankcase. When the compresser starts, the pressure is reduced and the oil may foam. Oil foaming will at the least cause poor lubrication; if the foam fills the crankcase, it can be pulled into the cylinder by the suction vapor. The noncompressible oil then causes hydraulic pounding, which can be (and often is) severe enough to break valves.

When the investigator hears the compresser turn on and off as it short-cycles, he should look at the sight glass to check whether the gas is low. He should also check the condenser's water-inlet and discharge temperatures or, on air-cooled condensers, determine if air is moving through. A quick reverse flush of the condenser or replacement of a fan belt may be all that is needed to get the chamber back into normal operating condition.

Gentner (1967) shows some typical malfunction symptoms for on-off direct-expansion systems (fig. 6.1). Similar data (fig. 6.2) for secondary-coolant and modulated direct-expansion systems show a number of similar symptoms.

Controls

The major problem with solid-state controls is a slow set-point drift due to instability of the bridge and amplifier circuit, a change in resistance in the circuit, or corroded or dirty sensors. Each system will have its own weak point, which can quickly be determined and used as a maintenance item. Resistance-element controls, for example, may suffer from a tendency for the sensors to break, or the rebalancing slide wire may cause such trouble that it has to be replaced as often as twice a year.

Spare Parts

Theoretically, the manufacturer of commercial, prefabricated growth chambers should be the best source of information regarding a proper selection of spare parts. Unfortunately some manufacturers know less about their products' maintenance problems than do the people who own them. The commercial chamber may have to be treated in the same way as those built by local contractors—namely, to rely on one's own and others' maintenance records to develop an appropriate spare-parts inventory. Since most of the parts needed can be obtained ''off the shelf''

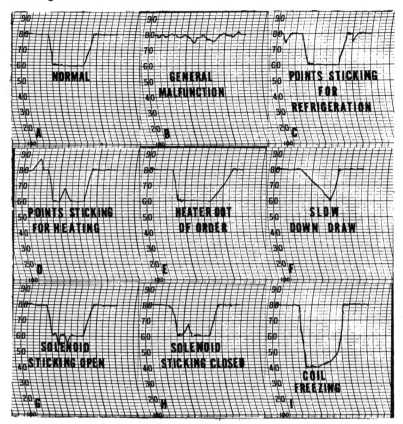

FIG. 6.1. Temperature progressions indicating various kinds of malfunctions in an on-off–controlled direct-expansion refrigeration system of a controlled-environment room (Gentner 1967).

from local commercial-refrigeration firms and air-conditioning-controls manufacturers, many items need not be stocked at all. Calling a manufacturer for assistance in troubleshooting a malfunction may or may not be helpful. As with any other mechanical problem, success would depend on getting in touch with the right person.

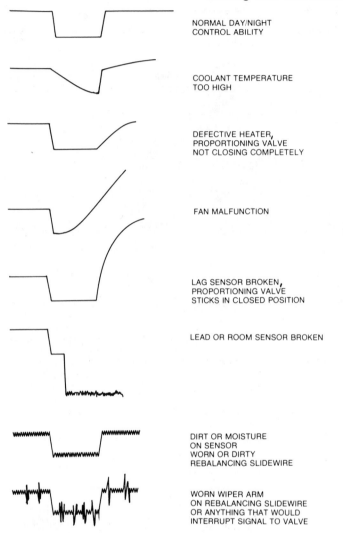

FIG. 6.2. Temperature progressions indicating various malfunctions in cooling systems for plant-growth chambers.

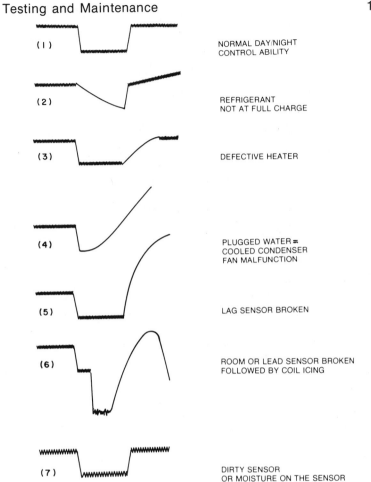

(1)   NORMAL DAY/NIGHT
       CONTROL ABILITY

(2)   REFRIGERANT
       NOT AT FULL CHARGE

(3)   DEFECTIVE HEATER

(4)   PLUGGED WATER =
       COOLED CONDENSER
       FAN MALFUNCTION

(5)   LAG SENSOR BROKEN

(6)   ROOM OR LEAD SENSOR BROKEN
       FOLLOWED BY COIL ICING

(7)   DIRTY SENSOR
       OR MOISTURE ON THE SENSOR

FIG. 6.2. (cont.) *Left,* electronic proportional control, secondary coolant; *right,* electronic, two-step sequencing control, direct expansion.

References

Gentner, W. A. 1967. Maintenance and use of controlled environment rooms. *Weeds* 15:312–16.

Price, S. G. 1970. *Air Conditioning for Building Engineers and Managers.* New York: Industrial Press.

Service Pointers. 1960. E. I. duPont de Nemours and Co., Freon Products Division, Wilmington, Delaware.

APPENDICES

APPENDIX 1

Conversion Factors

To convert from	To	Multiply by
BTU/ft²	cal/cm²	271×10^{-3}
BTU/ft³	kcal/m³	8.9
BTU/hr	kcal/hr	252×10^{-3}
BTU/hr	ton of refrigeration	8.33×10^{-5}
BTU/hr × ft² × °F	kcal/hr × m² × °C	4.88
BTU/hr × ft² × (°F/ft)	kcal/hr × m² × (°C/m)	1.49
BTU/lb	cal/g	555×10^{-3}
BTU/lb × °F	cal/g × °C	1
Cu ft/min, or cfm	m³/min	28.3×10^{-3}
Cu ft/min	l./min	28.3
Ft/min, or fpm	m/min, or mpm	304.8×10^{-3}
Footcandle, or ft-c	meter candle, or lux	10.76
Gallons/min, or gpm	l./min	3.78
Horsepower (electric)	BTU/hr	2547
Langley, or L	cal/cm²	1
Langley/min	mw/cm²	69.7
Mile/hr, or mph	m/min, or mpm	26.8
Pound, or lb	g	453
Pound/in.²	mm of Hg	51.7
Pound/in.²	g/cm²	70.3
Pound/ft²	kg/m²	4.88
Pound/ft³	kg/m³	16
Sq ft	m²	929×10^{-4}
Ton of refrigeration	BTU/hr	12,000
Ton of refrigeration	cal/hr	271×10^{-3}
Watt/cm²	cal/cm² × min	14.34
Watt/cm²	Joule/sec	1

APPENDIX 2

Useful constants for water and air

Water

1 gal weighs 8.345 lb
1 ft^3 weighs 62.428 lb
Specific heat = 1 BTU/lb × °F   or   1 cal/g × °C
Thermal conductivity = 0.330 BTU/hr × ft^2 × (°F/ft) or 0.5 kcal/hr × m^2 × (°C/m)
Heat of vaporization = 973 BTU/lb or 540 cal/g
Heat content of 50% RH vapor at 75 °F (24 °C), minus the heat content of water
at 50 °F (10 °C) = 1076 BTU/lb or 597 cal/g

Air

Standard air is dry air at 70 °F and 14.7 psi (20.5 °C and 760 mm of Hg)
Volume of 1 lb = 13.34 ft^3 or 378 l.
Volume of 1 kg = 834 l.
Density = 0.075 lb/ft^3 or   1.2 kg/m^3
Specific heat = 0.240 BTU/lb × °F   or   0.240 cal/g × °C

APPENDIX 3

Manufacturers of Growth Chambers in the United States

Controlled Environments Inc., P.O. Box 347, 601 Stutsman St., Pembina, N.Dak. 58271

Environaire System Inc., 141 Prospect St., East Longmeadow, Mass. 01028

Environator Corp., 2024 Gibson Dr., Warren, Mich. 48089

Environmental Growth Chambers, P.O. Box 407, Chagrin Falls, Ohio 44022

Forma Scientific Inc., P.O. Box 649, Marietta, Ohio 45750

Hotpack Corp., Hotpack Growth Chambers, 5086 Cottman Ave. and Melrose St., Philadelphia, Pa. 19135

Instrumentation Specialties Co. Inc. (I.S.C.O.), Building 978, Lincoln Park West, Lincoln, Neb. 68524

Lab-Line Instruments Inc., Lab-Line Plaza, 15th and Bloomingdale Aves., Melrose Park, Ill. 60160

National Appliance Co., P.O. Box 23008, 10855 S.W. Greenburg Rd., Portland, Ore. 97223

Parce Engineering Co., P.O. Box 2366, 900 W. Van Buren, Harlingen, Tex. 78550

Percival Manufacturing Co., P.O. Box 249, 1805 East 4th St., Boone, Iowa 50036

Scientific Systems Corp., 9020 S. Choctaw, Baton Rouge, La. 70815

Sherer Dual Jet Division, Kysor Industrial Corp., Marshall, Mich. 49068

Note: While it is impracticable to provide a complete list of manufacturers and dealers, this partial list is furnished for information, with the understanding that no discrimination is intended and no guarantee or reliability implied. The list was originally prepared by the Phyto-Engineering Laboratory, U.S. Department of Agriculture, March 1972.

APPENDIX 4

Manufacturers of Growth Chambers outside the United States

Austria

Ruthner Industrieanlagen fur Pflanzenbau Gesellschaft, M.B.H., Wein 2 Obere Donaustrasse 49–51.

Canada

Controlled Environments Inc., 611 Madison St. Winnipeg 21.

England

Climair Air Conditioning Ltd., 60 George St., Richmond, Surrey.
Environment and Air Conditioning Ltd., Mowbray Dr., Blackpool, Lancashire FY3 7UN.
Fisons Scientific Apparatus Ltd., Bishop Meadow Rd., Loughborough, Leicestershire.
R. W. Gunson, 20–21 St. Dunstan's Hill, London, E.C. 3.
Prestcold Ltd., Theale, Nr. Reading, Berkshire.

Japan

Koito Industries Ltd., Environment Control Division, No. 30, 1-Chome, Ohmiyamae Suginami-ku, Tokyo.

West Germany

Brown Boveri-York, Kalte- und Klimatechnik GmbH 68 Mannheim 1, Postfach 346.
Ernst Votsch, Kalte- und Klimatechnik Kg 7462 Frommern Wurtt, Postfach 40.
Karl Weiss Giessen, Fabrik Elektro-Physikal Gerate, D-6301 Lindenstruth.

Note: While it is impracticable to provide a complete list of manufacturers and dealers, this partial list is furnished for information, with the understanding that no discrimination is intended, and no guarantee or reliability implied. The list was originally prepared by the Phyto-Engineering Laboratory, U.S. Department of Agriculture, March 1972.

APPENDIX 5

Types of systems for measuring radiant energy that we have used or are familiar with

AGRICULTURAL SPECIALTIES CO.
11313 Frederick Ave.
Beltsville, Md.
Spectroradiometer; fixed wavelength

AGROMET DATA SYSTEMS
307 Highgate Rd.
Ithaca, N.Y.
Dome solarimeter

E. G. & G. INC.
Electronics Products Div.
170 Brookline Ave.
Boston, Mass.
Illumination meters, photodiode type; radiometer; spectroradiometer; standardizing lamps

EPPLEY LABORATORY INC.
12 Sheffield Ave.
Newport, R.I.
Pyranometer; thermopiles

GAMMA SCIENTIFIC INC.
2165 Kurtz St.
San Diego, Calif.
Illumination photometer; photoemissive type; spectroradiometer; radiometer; standardizing lamp

GENERAL ELECTRIC CO.
Nela Park
Cleveland, Ohio
Illumination meter; foot-lambert meter

HEWLETT-PACKARD
P.O. Box 28234
Atlanta, Ga.
Radiant flux meter system

INSTRUMENT SPECIALTIES CO., INC.
Building 978, Lincoln Air Park West
Lincoln, Nebr. 68524
Spectroradiometer; standardizing lamp

INTERNATIONAL LIGHT INC.
Dexter Industrial Green
Newburyport, Mass.
Illumination meter; barrier, photodiode or emissive type; spectroradiometer; integrating photometer

LAMBDA INSTRUMENT CO.
4421 Superior St.
P.O. Box 4425
Lincoln, Nebr.
Illumination photometer; radiometer; PAR meter; far-red or 790 nm sensor

PHOTOVOLT INC.
1115 Broadway
New York, N.Y.
Illumination meters

WEATHERMEASURE CORP.
P.O. Box 41257
Sacramento, Calif.
Pyranometer; pyranograph

WESTON INSTRUMENTS INC.   YELLOW SPRINGS INSTRUMENT
Newark, N.J.   P.O. Box 279
Illumination meter; barrier   Yellow Springs, Ohio
layer type   Pyranometer; Kettering radiometer

Note: The note to Appendix 3 applies here also.

INDEX

Air flow
  aspirated housing, 92
  aspirated psychrometer, 94
  controlled-environment room, 32, 87, 143
  direction, 88, 143, 144
  lamp loft, 66, 67, 147
Anemometer, 87–89
Aspirated housing, 45–46, 91

Barrier, 21, 64, 109, 146
Bailey, W. A., 62
Ballasts
  cooling, 72–73, 147–48
  life, 71–72
  remote placement, 73–74
Bellows thermostat, 40
Bimetal switch, 39–40
Biotron, 9
Borthwick, H. A., 8, 62, 108

Cabinets, plant growth, 5, 6
Carbon dioxide
  depletion, 81, 121
  enhancement, 117–22
  injection, 82–83, 144–45
Chemical driers, 52
Chili pepper, 114, 127
Cold plate, 51
Cold rooms, 5
Color temperature, 58, 60, 101
Compressor, 15
Condenser
  air-cooled, 16–17, 159
  water-cooled, 17–19, 159
Conduction, 21, 23, 28
Cooling coil, 12, 14
Cooling load
  barrier conduction, 21
  latent heat, 21
  make-up air, 22, 81

radiant energy, 21
wall transmission, 20
Corn, 116, 117, 123
Cosine law, 98–100
Cotton, 119
Cucumber, 118

Dark rooms, 5
Dehumidification, 51–52
Design paramaters, 9–10
Dew chambers, 5, 24
Direct-expansion systems, 13, 38–39, 142
Dogwood, 126

Evaporative cooling, 18–19, 29–31
Evaporator, 12, 14
Expansion value, 14

Flax, 126
Fluorescent lamps
  cooling, 65, 66, 146–47
  light output, 63–64
  lumen maintenance, 75–76
  mercury condensation point, 66
  physical characteristics, 62–63
  spectral energy distribution, 61, 63, 70
  tube-wall temperature, 64, 67
Foot candle, 122

Geranium, 117
Germinators, 5, 23–24
Gloxinia, 118
Greenhouses
  air conditioning, 32–33
  daily light intensity, 27–28
  glazing, 26–27
  heating, 28–29
  lighting, 33–34
  orientation, 28
  refrigeration requirements, 32–33

Greenhouses (*continued*)
  shading, 32
  standards, 29
  structure, 9, 26
  ventilation and cooling, 29–31
Growth-chamber performance, 153
Guidelines, 11, 108, 135, 140

Heat transmission, 20–25
High energy reaction (HER), 122
High-intensity-discharge (HID) lamps, 33, 34, 57, 58, 69, 70, 76, 80
Hot-gas by-pass, 13, 38
Hot-gas defrost, 142
Humidification, 49–51
Humidistats
  dew point, 54–55, 95–96
  Dunmore, 54–55, 95
  hair type, 52–53, 94
  lithium chloride, 52, 54, 95
Hydraulic thermostats, 40–42
Hyoscyamus niger, 124

Illuminance, 97–101
Incandescent lamps, 76–78
Inoculation chambers, 24–25
International System of Units, 97
Inverse square law, 98–99
Irradiance, 101–4

Jeffrey pine, 114

Lamp life, 76, 78
Lamps
  fluorescent, 61–67, 146–47
  incandescent, 76, 77
  Lucalox, 33, 57, 68, 69
  metal halide, 57, 76
  sodium, 57–58, 76
  xenon, 58, 59
Latent heat, 21, 22, 33
Lead-lag sensors, 46
Lettuce, 117, 127
Light
  measurement, 97–101
  photoperiod, 33
  programming, 74–75
  saturation, 56–57

spectral distribution, 57–59
  supplemental, 33–34
Loading of plant benches, 138
Lucalox lamps, 57, 68, 69
Lumen maintenance, 75–76

Maintenance, 155–58
Make-up air, 22, 81
Marigold, 117, 130
Mercury vapor pressure, 63, 67, 69
Metal halide lamps, 57, 76
Mist chamber, 25

Nutrient solution, 128

Oats, 125
Off-normal alarms, 148

PAR (photosynthetically active radiation), 56, 57, 102, 105, 121
Peanut, 130
Petunia, 119
Photoemissive transducer, 100
Photomorphogenic pigment, 62
Photoperiod, 33, 78, 126–27
Photoperiod rooms, 5, 8
Photosynthetically active radiation, 56, 102, 105, 121
Photovoltaic cell, 97–99
Phytochrome, 122–24, 126
Phytotron, 6–9, 34, 49, 66, 82, 107–8, 129, 135
Pinto bean, 118, 127
Plant-growth cabinets, 5, 6
Potato, 127
Power Groove (PG-17) lamp, 64, 65, 66, 75
Pressure reducer, 12
Proportioning valves, 38
Pull-down tonnage, 22
Pyranometer, 102

Radiant energy, *see* Light
Radiant temperature, 92, 122
Radiometer, 92
Radish, 118, 119, 120, 121
Red–far red ratio, 62, 76, 78, 103, 104–5, 123, 126
Red fir, 114

Reflectance, 78–80
Refrigeration
  calculation of load, 19–22
  cycle, 12, 13, 19
  size, 19–22
  system, 150–51, 159–60
Relative humidity, 47–56
Resistance thermometer, 42–43
Response time, 40–41
Reverse air flow, 88
Roomettes, 5, 7

Salvia, 129
Secondary coolant, 38
Seed germinators, 23–24
Sodium lamp, 57–58, 76
Solar radiation, 56, 59, 154
Solivap Green, 32
Soybean, 119
Specification guidelines, 11
Spectral energy distribution, 57–63
Spectral irradiance, 105–6
Spectroradiometer, 105–6
Specular aluminum, 79
Substrate, 129–30
Sugar beet, 123

Temperature
  controllers, 39–47
  cooling coil, 37–38

growth, 113–15
range, 141–42
seasonal progression, 115
set point, 37
Thermal load, 143
Thermistors, 44
Thermocouples, 45–46, 90, 93
Thermoperiod, 114
Thermopile, 101–2
Thermostats
  bimetal, 39–40
  gas-filled, 40
  hydraulic, 40–42
  location, 45–47
  resistance thermometers, 42–43
  thermistors, 44
  thermocouples, 45
Tobacco, 113, 117, 118, 121, 127
Tomato, 115, 116, 117, 118, 119, 125, 127

U value, 20, 29

Vapor pressure, 48
Vapor-pressure deficit, 49

Walls
  heat transmission, 20–21, 139
  reflectance, 78–80, 139

Xenon arc lamp, 58, 59

97223

DATE DUE

LIBRARY